D1573310

UNFINISHED VOYAGES

Western Australian Shipwrecks 1851–1880

UNFINISHED VOYAGES

Western Australian Shipwrecks
1851–1880

GRAEME and KANDY-JANE HENDERSON

UNIVERSITY OF WESTERN AUSTRALIA PRESS

First published in 1988
by the University of Western Australia Press. Nedlands W.A. 6009

Agents: Eastern States of Australia, New Zealand and Papua New Guinea: Melbourne University Press, Carlton South, Vic. 3053; U.K., Europe, Africa and Middle East: Peter Moore, P.O. Box 66, 200a Perne Road, Cambridge CB1 3PD, England; U.S.A., Canada and the Caribbean: International Specialized Book Services Inc., 5602 N.E. Hassalo Street, Portland, Oregon 97213, U.S.A.; Singapore and Malaysia: National University of Singapore Multi-Purpose Co-operative Society Ltd, Ground Floor, Central Library, Kent Ridge, Singapore 0511.

Designed and photoset by University of Western Australia Press.
Printed by Kyodo-Shing Loong Printing Industries Pte Ltd, Singapore.

National Library of Australia
Cataloguing-in-Publication data

Henderson, Graeme, 1947- .
Unfinished Voyages—Western Australian shipwrecks
1851-1880.

Bibliography
includes index. ISBN 0 85564 282 3

1. Shipwrecks—Western Australia—History—
19th century. I. Henderson, Kandy-Jane, 1955- .
II. Title.

994.1

Contents:

List of Illustrations

Acknowledgements

This book, like its predecessor, evolved from the catalogue of shipwrecks at the Western Australian Museum. We thank John Bannister and Dr Ian Crawford for their support. The Battye Library provided the major sources of information for this volume. We thank the staff of the Battye Library for their assistance, and especially note the support of Margaret Medcalf and Chris Coggin.

The Literature Board of the Australia Council supported the writing of the first volume, *Unfinished Voyages 1622–1850*, with a grant in 1976, and supported this second volume with a Special Purpose Grant in 1985, enabling travel to archives in Sydney, Melbourne and Canberra. Among the archival institutions which have been particularly helpful are the British National Maritime Museum, the Public Record Office in London, the State Archives of Tasmania, the Public Record Office of South Australia, the National Library of Australia, the Mitchell Library, the La Trobe Library, the Department of Transport in Canberra, and the Australian Archives staff in Canberra, Sydney, Melbourne and Adelaide.

Associate Professor Frank Broeze, who lectures in Maritime History at the University of Western Australia, again read the drafts and gave many valuable suggestions. The drafts were read and commented upon by Dr Ian Crawford, Sally May and Mike McCarthy of the Western Australian Museum.

Pat Baker again supplied his talents for photographic work.

The staff of the University of Western Australia Press—particularly Vic Greaves, who has persistently encouraged the production of a second volume, and Susannah Crow—were very supportive.

Among the many others who helped in supplying information regarding particular entries are, Harold Roberts, Bruce Melrose, Andrew David, David Lyon, Mike Lorimer, Mark Staniforth, Maritime Archaeology Association of Western Australia members (Mike Pollard, Richard McKenna, Denis Robinson and David Totty), Scott Sledge, Charles Staples, Jill Worsley, Professor Geoffrey Bolton, Eric Grose, David Hutchison, Graham Anderton, Warren Robinson, Jim Henderson, Shirley Hands, Sister Albertus and Dr John Williams.

The Department of Arts, Sport, Environment, Tourism and Territories, responsible for administering the Historic Shipwrecks Act, assisted through its fieldwork grants to the Western Australian Museum.

Susan Cox typed the drafts.

The views expressed are our own and not necessarily those of the Trustees of the Western Australian Museum or the Library Board of Western Australia.

Our thanks to Amy and James Grant for their support.

Graeme and Kandy-Jane Henderson

Introduction

Western Australia became a coast of shipwrecks during the second half of the nineteenth century. Publicity about seventeenth century Dutch shipwrecks has given rise to a popular notion that the Western Australian coast is littered with an extraordinary number of ancient shipwrecks. Yet, despite the fact that the coast lay astride a major sea route from Europe to Asia, the records indicate that only ten substantial vessels were lost in these waters prior to the establishment of a colony at the Swan River in 1829. The settlement brought shipping inshore to more dangerous waters, and in these early pioneer years some of the newcomers were wrecked in the unfamiliar surroundings. But the statistics now available show that it was later in the nineteenth century, when the volume of shipping trading with ports on the Western Australian coast expanded, that the numbers of shipwrecks increased. During the ten years from 1829 to 1839, for example, about ten substantial vessels were wrecked, whereas some eighty were wrecked between 1870 and 1880.

The Colony changed considerably during the years 1851 to 1880 as the experiment of convict transportation was followed by a period of adjustment. During the 1850s and 1860s the overseas shipping movements of Fremantle were modest, and the umbilical cord with Britain remained the most important trade route, imports from Britain never falling below 70 per cent of total imports.

The major trades gradually developed in volume. Wool continued as the most important export, but timber and pearling industry products were also substantial by 1880. Food and beverages, and clothing, were the most important groups of imports. The United Kingdom remained the dominant trading partner, involved in around half the trade at any one time.

Two other significant trade regions were the Australian colonies and New Zealand, and Afro-Asia and the Indian Ocean. Of these, the Australian colonies trade achieved ascendancy by 1880, with 36 per cent of the total. However, much of this trade consisted of re-exports—a channelling of British trade with Western Australia through the Eastern Colonies. Because of Western Australia's isolation, shipping services to a large extent implied the direction of trade and thus imposed parameters of economic development.

The pattern of British-Fremantle shipping in 1850 was largely trilateral, most of the tonnage returning to London via Afro-Asian ports to gather return cargoes. By the late 1860s, the wool industry had progressed to the extent that virtually all arrivals in late winter and early spring were assured a full homeward cargo, and these vessels followed a terminal route. By 1880, this pattern was supplemented by a strong easterly flow of traffic from Britain through Fremantle to the Eastern Colonies and Asia.

Fremantle's shipping links with Asia changed less dramatically than those with Britain over the period. However, the volume of shipping moving between Fremantle and the Eastern Colonies increased at a faster rate than that with either Britain or Asia. In the 1870s, the most significant shipping development was the introduction of a regular steam service with Adelaide.

The pattern of coastal trading was transformed by the arrival of steam services in the 1870s. Previously, some coasting had been done by the wool ships from London, or the timber ships from the Eastern Colonies, but the small locally owned coasters were never able to provide an adequate service, particularly around Cape Leeuwin to link up with Albany, where the European mail steamers called.

Ownership of the vessels engaged in trade with Britain was fairly clear cut. Frederick and Charles Mangles were involved from the mid 1830s. The big London shipowner Duncan Dunbar became interested with the commencement of convict transportation. By the 1870s, Wilson and Co., of London, owned a substantial majority of the regular vessels. Robert Habgood, of London, also owned two regular vessels. Western Australian merchants Walter Padbury and John Bateman owned one each. The Felgate Line of London, involved in the Fremantle trade since 1840, ran Wilson's vessels in the trade. The establishment of a substantial locally-owned company for the London trade was the subject of perennial discussion in the Colony, but few of the projects got off the ground and none ever attained the necessary magnitude to carry out the envisaged task adequately. In around 1877, the British shipowners and brokers engaged in the trade between Britain and Australia formed themselves into a loosely organised body known as the Associated Australian Owners and Brokers to maintain a regulated outward loading schedule.

The population of Western Australia had increased only gradually, reaching nearly 30,000 by 1880. By that time, economic resources were being exploited along almost the entire coastline, and on the offshore islands as well. To service these industries, more vessels were required. In 1870, the boatbuilding industry at Fremantle was beginning to expand to cater for the pearl fishery in the north.

Larger and more technologically advanced vessels were introduced to the overseas trades. In 1850, all the sailing vessels arriving at Fremantle were built of wood, but by 1880, there were composite vessels and iron vessels. The large convict ships stopped coming in 1868, but the cargo vessels arriving from Britain increased in size to between 300 and 400 tons. The vessels employed in trade with Afro-Asian ports and the Eastern Colonies were generally smaller, varying between 100 and 250 tons. Among these vessels wood construction predominated to a greater degree, and although a few larger vessels were composites, hardly any were built of iron. Their quality was generally inferior to those employed in the London trade. The coastal trade vessels were smaller again, and, because of their size, they were slow, and could easily be overwhelmed by heavy seas. With the arrival of steam services on the coast the local shipbuilders were no longer called upon to construct coastal trading vessels in much quantity.

Apart from the overseas mail steamers calling at Albany, steam was not introduced to Western Australia until the mid-1870s. By that time, the steamship was a well advanced engineering product, so Western Australia did not see in any numbers the wooden paddle-steamers, powered by crude beam engines, that were common in Europe in the earlier part of the century.

During this period, the numbers of shipwrecks increased dramatically, and the losses occurred over a greater portion of the coastline than hitherto. Small wooden vessels continued to be wrecked, but so too were a number of large iron ships, including some steamers.

The Colonial administration pondered over the reasons for the growing number of casualties, and introduced some measures to address the problem. Britain's Merchant Shipping Act only applied to Western Australian-registered ships when out of the jurisdiction of the Government of Western Australia. Local ordinances, however, began to require reasonable standards. An ordinance of 1861 required that ships be surveyed for seaworthiness before clearing out from port. One ordinance of 1863 limited the number of passengers carried to one person for every 2 tons of the registered tonnage. Another, of 1875, required that no passenger ship go to sea without having on board a chronometer in proper working order. The dearth of navigational aids had been a frequent source of complaint since the beginning of settlement. Many of the worst disasters occurred on dark nights when lighted marks, had they existed, might well have guided navigators to safety. The need for better charts resulted in the creation, in 1872, of a joint Admiralty and Colonial survey unit, and in 1880, a 125-ton schooner was acquired to replace the whale boats previously used by hydrographic surveyors. More ports were surveyed, but port facilities were slow to appear in these pre-gold-discovery days, when money for capital works was very limited.

The lack of a harbour at Fremantle was particularly irritating to local residents. For this reason, the overseas mail steamers continued to make Albany their only Western Australian port of call, thus depriving Fremantle of any communications, migration and trading benefits. The requirement for shipping to anchor on Gage Roads created a thriving lighter and river trade at Fremantle.

The increasing number of shipwrecks also highlighted the need for closer control over crews, as well as over the ships themselves. To improve the standard of efficiency of navigators and crews, and to better ensure the safety of their vessels, comprehensive legislation and regulations were brought into effect, and these were more rigorously enforced. An ordinance of 1868 required that coasting vessels be commanded by masters certified by the Board of Trade. When vessels were lost, new procedures assisted in ascertaining the cause of the loss, and in laying blame.

In recent years, there has been a growing community and government awareness of the importance of shipwrecks in the second half of the nineteenth century. These form a cultural resource that can be used for scientific study and recreation by both present and future generations. This book lists every known shipwreck and major casualty that occurred in the waters off Western Australia during the years 1851–1880.

The structure of this volume is similar to that of its predecessor, which covered the period 1622–1850. A register of sites, both found and unfound, has been clothed with enough information about the history of this region to provide reading for a general audience as well as a reference volume for researchers and archaeological site managers. It is intended that the book make more people aware of the archaeological, historical and recreational resources lying beneath our waters. It should also stimulate further research in the areas of maritime history and archaeology by providing a source book with a comprehensive listing of sites.

Entries include the name, official number and date of each casualty, together with an account of the circumstances of the loss, and a background of the vessel's construction and career. Significant strandings are included, but in shortened format. Some details are given about where vessels were thought to have been lost, and whether the wreck sites have been located in modern times. Accurate locations are given for gazetted sites. We have provided maps showing all wrecks on every part of the Western Australian coast up to the year 1880. The names of located wrecks are given in capital letters on those maps.

It is apparent that shipwreck sites are often found by people who are not aware of the significance of their find, and it is hoped that this book will make it easier for divers to decide whether the site they have located was previously unknown. Two examples can be cited where the first volume of *Unfinished Voyages* succeeded in this way. An unidentified early whaler site (the earliest known on the Western Australian coast) on the remote Rowley Shoals was reported to the Western Australian Museum because the finder learnt that the earlier volume cited archival evidence of an early wreck there, and realized that he had found something of significance. On another occasion, the finders of a site near Cape Leeuwin read the *Cumberland* (1830) entry and, realising that it was likely they had indeed found the remains of this ship, reported it to the Museum. Examples of rewards given for the reporting of particular sites will be found among the entries. It should be borne in mind that some of the smaller rewards were paid some years ago.

Pilot's Boat, 1851

Port Pilot Edward Back

At the beginning of 1851, the Fremantle pilot's normally arduous job had its compensations. Rottnest Island was better situated than Fremantle for sighting approaching vessels, so the pilot had a house on Rottnest, and the fair isle afforded the opportunity to occasionally mix business with pleasure.

Edward Back, the Port Pilot, left Fremantle in the pilot boat on the morning of 14 January for Rottnest Island, with three women on board as passengers. On the way, Back intended to board the American whaler *Gentleman*, which had just arrived from the whaling grounds. A heavy sea was running, so the captain offered to hoist the pilot boat aboard with the passengers still in it. But a heavy wave struck when the boat was about half-way up and she broke in two, throwing the women into the water.[1] (See map on p. 242.)

The ship's boat rescued the humiliated women from their perilous position without any other harm than the fright occasioned by their involuntary bath.[2] All the oars and the mast were picked up, but the sail was lost and the hull, apparently rotten anyway, was deemed to be not worth the expense of another repair.

It was not the first time that Back had blotted his copybook, and Governor Charles Fitzgerald, irritated that the craft had been used as 'a passage boat for females', suggested that Back should have to pay for a replacement boat, and refused the boat crew's request for replacement of their lost clothes.[3] Later, the Governor relented on the clothes issue, each man being given one pair of trousers, one pair of shoes, one shirt and one cap. But they lost the use of the Rottnest house, and were given the loft of the government boat-house in Fremantle for accommodation.[4]

The pilot's boat was described as a whale boat. It was probably a double-ender of around 9 metres in length, and similar to those used by the bay whalers at Fremantle from the 1840s.

NOTES

1. Daniel Scott (Harbour Master) to Col. Sec., 14 January 1851, C.S.R. 219, fol. 266.
2. *Perth Gazette*, 17 January 1851.
3. Charles Fitzgerald (Governor) to Col. Sec., 18 January 1851, C.S.R. 219, fol. 266.
4. Gov. to Col. Sec., 25 January 1851, C.S.R. 219, fol. 269.

The King George Sound pilot boat in the 1850s — a whale boat like that of Pilot Back.

Venus

The master of the tiny schooner *Venus* had problems in crossing the starting line on what was to be the vessel's final voyage in September 1850. Captain Mason and his crew of four attempted to get the vessel under way from Fremantle on the 18th, but returned to anchor at 4 p.m. because it was too calm to make headway. The following day, they set out again, but this time an afternoon gale forced them to return to shelter. The *Venus* proceeded without interruption towards Batavia and Singapore on the 20th, with a cargo of 20 tonnes of sandalwood destined for Chinese joss-houses, via the middlemen in Singapore.[1]

In the Straits of Singapore on 25 November, the *Venus* had a brush with pirates, who chased the schooner for 6 hours. But this time, the weather turned to the favour of the *Venus*—a fresh breeze sprang up and enabled her to escape.

Captain Mason completed his trading and headed back towards the Swan River with a new cargo and the European overland mail. Unfortunately, at 11 p.m. on 10 April 1851, the *Venus* struck one of the reefs surrounding the Southern Group of the Abrolhos Islands and was wrecked.[2] The *Venus*'s men had endured a passage of 10 weeks from Java Head with insufficient water, so they were severely emaciated when the vessel struck.[3] At daylight, the crew and the one passenger started out in the small dinghy to seek water, leaving John Williams, the cook, on board the *Venus* because he was too weak to want to be moved.

They rowed across to Middle Island. The *Venus* had made several visits to these islands before, the first being in 1842, when the vessel had carried a salvage party from Fremantle to the wreck of the *Ocean Queen*. So it is probable that the crew knew that there were freshwater wells on Middle Island.

Once ashore, the men were too emaciated to row back to the wreck of the *Venus* for 5 days. When they did return, Williams was still on board, but he again declined to accompany them to land, saying he was too weak. The Middle Island men, who had been living on shellfish from the rocks, took sugar and rice from the galley of the *Venus*.

A week later, they went to the wreck again. Williams was still alive and said he would return to shore with them if they would carry him, but they were too

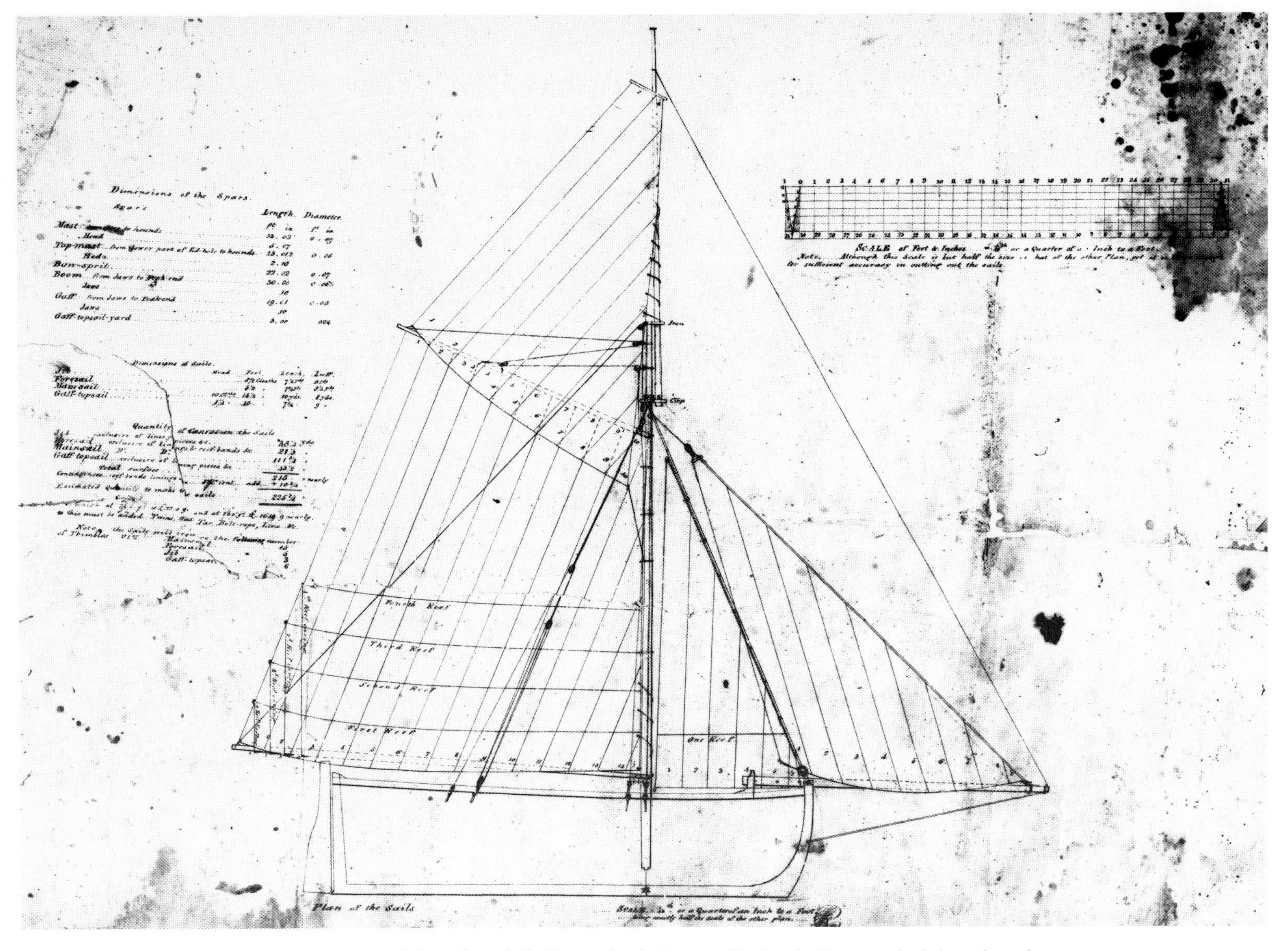

The lines of an early nineteenth century cutter, passed down through the Thomas family. It is possible that the *Venus* was built from these plans.

exhausted to do so. Three days later, they returned a third time and found Williams dead in his berth.

When Captain Mason had recovered sufficiently after the wreck, he rigged and decked the dinghy from the *Venus*. Meanwhile, he and his companions subsisted mainly on the meat of seals which they killed on the surrounding islands. Then on 15 October, Mason set out for Champion Bay on the mainland, accompanied by the passenger and one of the sailors. Mason had made a second dinghy during his enforced stay, and he left this with the two sailors remaining on the island, in case of any disaster to himself.

Mason's party reached Champion Bay on the night of the 15th, and delivered the European mail to the Postmaster. A week later, the schooner *Evergreen* called at the Bay and was sent to rescue the remaining sailors from Middle Island.[4]

The exact position of the *Venus* wreck is not known. It may be presumed that the vessel was heading southward when she struck, and that the wreck lay in very shallow, calm water on the Southern Abrolhos. Otherwise the wreck would not have held together sufficiently for Williams to lie in his bunk for 2 weeks before he died, and the weary crew could not have rowed between Middle Island and the wreck as frequently as they did. (See map on p. 44.)

In 1879, John Forrest, then the Colony's Deputy Surveyor General, visited the Abrolhos and recorded having seen on Middle Island the remains of two stone houses erected by a fishing party from Fremantle. Another visitor, on the cutter *Victoria* in 1867, had seen the frame of a wooden house there.[5] The Pelsart Fishing Company, formed in 1847, had been given permission to erect buildings on the Abrolhos and it seems likely that they would have selected Middle Island for the base of their operations, because of the fresh water. In 1848, the *Venus* had visited the facilities of the Company, which was curing bêche-de-mer and mining guano.[6] So the *Venus*'s survivors probably enjoyed at least some of the comforts of home during their stay on the island.

During Forrest's visit to Middle Island, he saw a grave with a headstone inscribed:

> Here lieth the body of John Williams Seaman, died April 1851 in the wreck of the *Venus* aged 41 years.[7]

Captain John Thomas

In 1978, an expedition from the Western Australian Museum, which was investigating the wreck of the *Zeewijk*, found an unmarked grave on Middle Island, located 300 metres north of two stone structures.[8] A police officer removed the skeleton and took it to Perth for analysis, which indicated it to be an Asian man rather than a Caucasian.

The 21-ton *Venus* was built on the Swan River, probably close to Perth, in 1839 by Captain John Thomas, who was building and repairing boats from the time of his arrival in the Colony in 1829. The *Venus* was constructed of jarrah timber and was not initially copper sheathed; it was only to please the insurers in Singapore that Thomas allowed her to be coppered there three years later. She then ran without any repairs until 1849. During this period, the *Venus*, generally under the command of Captain Thomas, voyaged twice to Singapore and visited Batavia and Adelaide, as well as being employed in coasting work.

The owner of the *Venus* was variously designated as Thomas, J. Thomas and Co., and Thomas Brothers. The principal appears to have been John Thomas, whose father John and brother James had been convicted, in 1835, of plundering the wreck of the *Cumberland* at Point Peron.

The *Venus* was originally 10 metres long by 3.7 metres in beam by 1.9 metres deep, and was iron fastened and carvel built with a square stern.[9] She was built as a cutter, with one mast and one deck. However, in August 1850, just prior to her last voyage, she was lengthened and refitted to measure 28 tons. She gained a mast and became a schooner with a standing bowsprit and scroll head.[10] At that time, Thomas later reflected, her timbers and planks were as sound and fresh as when she was built.[11]

NOTES

1. *Inquirer*, 25 September 1850.
2. *Perth Gazette*, 6 December 1850.
3. *Inquirer*, 12 December 1851.
4. *Inquirer*, 19 November 1851.

5. *Inquirer*, 23 October 1867, Supplement.
6. *Western Australian Almanac*, 1849, p. 27. See also *Perth Gazette*, 13 April 1848.
7. John Forrest, Report on the Examination of Houtman's Abrolhos for Guano Deposits, 10 April 1879, Lands and Surveys Department Inward Correspondence from Col. Sec., Acc. 223, Battye Library.
8. This fieldwork was financed by the Australian Research Grants Committee. The principal investigators under the grant were J. Green, G. Henderson, and C. Ingleman-Sundberg.
9. Register of Arrivals at Fremantle.
10. Register of Arrivals at Fremantle.
11. *Inquirer*, 13 September 1871.

Black Swan

The 14-ton cutter *Black Swan* was employed by the local shipowner Anthony Curtis between Fremantle and Castle Rock to the south of Busselton. He used the vessel to ship cattle from the farms of settlers around Cape Naturaliste to market at Fremantle. The *Black Swan* was reported as arriving at Busselton in October 1843 under the command of a Mr Leary (perhaps Edward Leary, a labourer), with a crew of three.[1] An advertisement in the *Inquirer* of January 1850 read as follows:

> For Bunbury and Port Vasse. To sail on 12 February, and on each succeeding month (weather permitting). The fine cutter *Black Swan*. For freight for Bunbury 30/–, Vasse 40/– per ton. This charge will cover carting from the River Jetty and storage until shipment.[2]

In May 1851, the *Perth Gazette* reported the loss of the vessel:

> A day or two before the commencement of the gale last week, as Mr Curtis' cutter the *Black Swan* was working out of the Murray Estuary, she unfortunately got aground on the bar, and the crew not being able to get her off before the gale, she became a total wreck.[3]

An 18-ton cutter named *Black Swan* was built at Westernport, New South Wales, for Alexander Livingstone of Newcastle. Its dimensions were 11 metres by 2.8 metres by 1.9 metres, and it had a running bowsprit, square stern, carvel build, no galleries and no figurehead. Unfortunately, the Register gives no indication of her fate beyond the word 'lost'.[4] (See map on p. 242.)

NOTES

1. C. Cammilleri, *Anthony Curtis: His Life in Western Australia* (Perth, 1963).
2. *Inquirer*, 30 January 1850.
3. *Perth Gazette*, 16 May 1851.
4. Register of British Ships, Sydney.

Unnamed Boat, 1851

The process of loading and discharging seagoing ships in anchorages exposed to winter gales at Fremantle was fraught with difficulty and danger. Lighters were frequently lost along with their cargo, and sometimes their crew. The situation was not satisfactorily resolved until the new harbour came into use in the 1890s.

On 9 May 1851, the *Perth Gazette* reported:

> We regret to hear that on Wednesday afternoon Mr Gray's 10 ton boat, which was loaded with a full cargo of tea and sugar worth nearly £400, which she had taken in from the *Empress*, and to which vessel she was hanging on in Owens Anchorage, was by the violence of the gale obliged to cast loose, and in a vain attempt to run into the South Bay was cast on shore, where she quickly went to pieces, none of the cargo being saved.[1]

Henry Gray, the shipowner and merchant, later wrote to the Editor to correct several particulars of this account:

> First you are told she had a full cargo on board, when in fact she had taken in not more than a quarter of a cargo the day before the gale came

> on, and was waiting for fine weather to complete it. Next you state the value of the cargo at nearly £400, whereas about a sixth part of that amount would pay for it. She was not also obliged to cast loose, the fact being that she parted 2 hawsers, one about 6 the other about 5 inches, and lastly she did not attempt to run into the south bay, as the wind was right ahead, and the strength of the gale so that it was impossible to show canvas to it, her anchors were let go in Owens Anchorage and she dragged them ashore.[2]

No further details of the incident are available. (See map on p. 242.)

NOTES

1. *Perth Gazette*, 9 May 1851
2. *Perth Gazette*, 16 May 1851.

Small craft in South Bay, Fremantle, in the 1850s.

Bee

The *Bee*, a small schooner, was driven ashore during a gale at the Vasse in October 1849. In May 1850, the vessel was refloated, but being inadequately moored, was again driven on shore.[1] Later that year, she was refloated and repaired, and in February 1851 was advertised for the coasting trade:

> The Colonial Built Schooner *Bee*, now lying at Fremantle, will in a few days be ready for sea, and will sail for Bunbury and the Vasse, or Champion Bay, if sufficient freight offers. For freight or passage apply to the owner, at Mr Lodge's, Fremantle, or Mr Turner, Perth, where goods will be taken in to be forwarded.[2]

Three months later, however, the *Bee* was a total wreck near Cape Bouvard, some 40 kilometres north of Bunbury.[3] The vessel was wrecked during boisterous weather, but no lives were lost. The *Bee* had on board 5 tonnes of flour shipped at Bunbury for the Commissariat at Fremantle, as well as 5.3 cubic metres of barley and 3 tonnes of potatoes. Nearly all the flour was recovered, as the salt water did not penetrate the sacks more than half a centimetre, which would appear to indicate that the *Bee* was wrecked ashore on the mainland.[4] At the time of her loss, the *Bee* was owned by a Mr Turner. (See map on p. 242.)

Nothing is known of her building or dimensions. However, if the owner was James Woodward Turner of Augusta, then it is possible that this vessel, like the *Alpha*, was built at Augusta.

NOTES

1. *Inquirer*, 15 May 1850.
2. *Perth Gazette*, 7 February 1851.
3. *Perth Gazette*, 17 May 1851.
4. *Perth Gazette* , 23 May 1851.

Unidentified Ship, Greenough River, 1851

In September 1851, the *Perth Gazette* renewed speculation about the fate of the *Mercury*, which disappeared while on a voyage from Calcutta to King George Sound in 1833. The *Gazette* reported:

> The [Champion] Bay natives have given us a very clear description of the loss of a large vessel with all hands, at the mouth of the Greenough, about 16 or 17 years ago, which corresponds to the time the *Mercury* was lost; there is nothing of her to be seen there now but the natives say a great quantity of things have been covered by the sand on the beach. This verifies the report brought in by the natives in 1834 . . .[1]

A further story was run the following month, after the return of the *Saucy Jack* from the north:

> Portions of a burnt wreck have recently been found a few miles south of the Greenough and the natives there detail the affair with great minuteness. All hands were drowned in consequence of the rush of water out of the mouth of the river preventing their swimming and it has been conjectured the vessel was the *Mercury* from Calcutta some years since. Numerous pieces of wreck have recently been picked up in the vicinity of Champion Bay, and a water butt apparently having been but a short time in the water.[2]

There were no further reports at that time about the wreckage (although there were more reports in 1861). It could have come from any one of a number of vessels. However, any wreckage found on shore in this area will be of great interest to maritime historians trying to solve the mystery of the *Mercury*.[3] (See map on p. 53.)

NOTES

1. *Perth Gazette*, 26 September 1851.
2. *Perth Gazette*, 17 October, 1851.
3. The *Mercury* (see *Unfinished Voyages 1622–1850*) carried the enterprising participants of an agricultural project intended for the south-west region.

Two Sons

The 16-ton cutter *Two Sons*, a small vessel built at Bunbury, was intended for the run between that port and Fremantle. In December 1851, she was driven on shore a little north of Bunbury, and although the cargo was recovered without much damage, the cutter was reported to be a wreck.[1] However, she was refloated and repaired, and in October 1853 was engaged in bringing oil up to Fremantle from Bunbury.[2]

The *Two Sons* was built by Benjamin Jackson (who also built the *Unknown* and the *Frolic*) and was initially owned by John Wenn. The vessel was 12.4 metres by 3.9 metres by 1.7 metres, and had one deck and a square stern. The same cutter was stranded at Fremantle in 1873.[3]

NOTES

1. *Perth Gazette*, 12 December 1851.
2. *Perth Gazette*, 14 October 1853.
3. Register of British Ships, Fremantle.

Eglinton

The barque *Eglinton* sailed from Gravesend bound for Swan River on 11 April 1852. She touched at Madeira and Tristan Da Cunha, where Captain Bennett found that his chronometer was giving incorrect readings. At the Cape of Good Hope, he took on more passengers and began the long eastward voyage across the Indian Ocean on 29 July, without bothering to correct the chronometer.

Five weeks after leaving the Cape, the crew noticed a discoloration of the water, indicating the proximity of land, and on Friday, 3 September, Captain Bennett told the passengers they would see Australia the following morning. He supposed himself to be 240 kilometres from the land, so he shortened sail to lessen the speed of the ship during the hours of darkness.

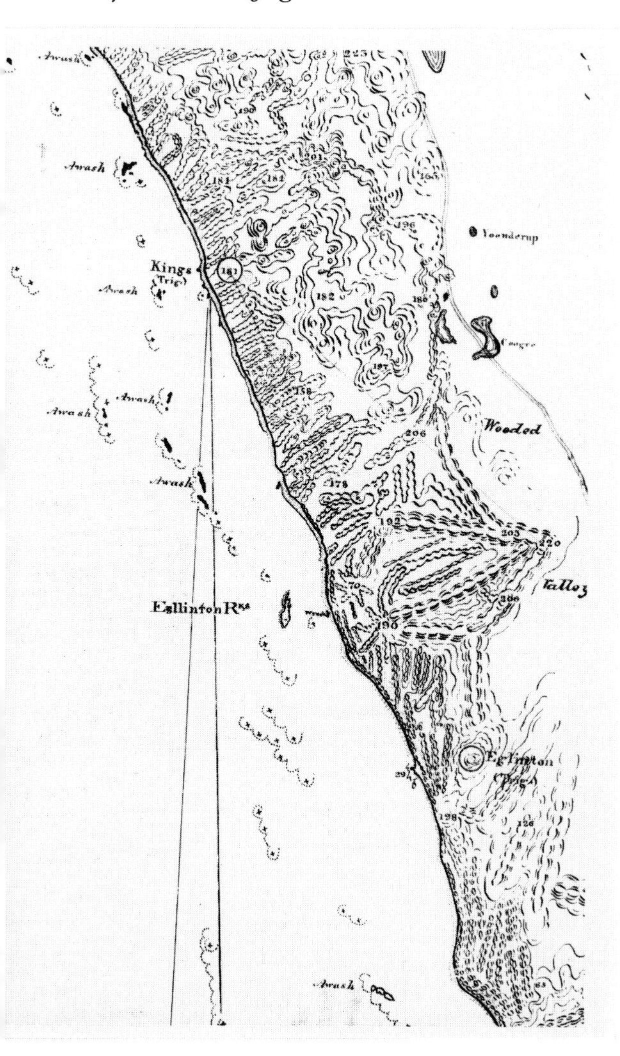

This 1875 chart, by Archdeacon, places the *Eglinton* in the wrong position, possibly because some wreckage was washed on shore.

The events of the following days are detailed in a letter from passenger Mrs James Walcott, a matron firmly in the mould of her times. The weather that evening was miserable, and Mrs Walcott went to bed at 9.30 p.m.:

> I lay still intending to undress afterwards, but the [supper] plate was only just placed by my side, when I was nearly jerked out; a crash accompanied by such a sound as I can never forget. I jumped up, thankful that I had my clothes on, but before I could stand came another and another—Oh God! We could hear her 'strong timbers' breaking up!—I recollect my first words were, 'don't scream'![1]

The *Eglinton* passed over the first reef and then drifted, rudderless, for a while before becoming fixed and helpless on the next reef. The barque was badly holed:

> The water was up to the seats and we were chilled! the night increased in severity: terribly dark and dreadful squalls of hail and rain and a fearful sea. A little tea was brought which we sipped and sat waiting for death.[2]

The masts were cut down and the boats made ready for launching. At first light, they were put in the water, and the task of saving the women commenced:

> I went up to the poop, held by Mr H. and another: the poor Captain was standing the image of silent woe! Pem [her son] was with me, suffering on my account; he begged me not to be frightened; in a few moments I should be safe in the boat; God bless you my dear Mother! He then held me, while the Captain tied the rope round my waist, my knees and feet; he then put the lifebuoy over my head, low enough for the arms to rest on, called out to the mate in the boat that all was ready. I stood without asking any questions, but when he placed me on the stern, overlooking the foaming waters, I asked what he was going to do. Drop you he said, it is the only way or chance![3]

The women were dragged the 20 metres to the ship's boat and taken ashore. Then it was the turn of the other passengers.

The survivors—all but two of the ship's complement were saved—were taken to Perth. A salvage party rescued much of the cargo, including 65,000 gold sovereigns for the Colonial Government. Captain Bennett was charged with negligence and convicted. He was fined a mere £50, and even this was paid by a collection among sympathetic colonists.

The wreck was re-located in 1971, at latitude 31°39′ south, longitude 115°40′ east, by spearfishermen Peter Boonman and Brian Castle, who were paid a reward of $600 by the Trustees of the Western Australian Museum that year. (See map on p. 22.) During two short seasons of excavation in 1971 and 1972 it was found that, although little of the vessel itself had survived on the seabed, a large quantity of the general cargo was scattered about the site. Among the items collected were hundreds of glass tumblers, wine goblets, other assorted glassware, sealed jars of preserved fruits, china dinner sets, druggists' goods, clay pipes, a toothbrush and a comb. Some of this material now forms a display at the Fremantle Museum.

The barque *Eglinton* was built in Quebec in 1848, but was soon bought by the London merchant John Jaffray. She was 36.2 metres long with an extreme breadth of 8.2 metres and a depth of hold of 5.5 metres.[4] The vessel carried a standing bowsprit with the figurehead of a woman, a square stern and no galleries. She had a poop deck in addition to a full deck.

A bottle of preserved fruit from the *Eglinton* wreck.

NOTES

1. Copy of a letter from Mrs James Walcott to her children at the Cape, 4 December 1852, Henderson Papers, Reel 133/1, Norfolk Record Office, England.
2. Walcott to children.
3. Walcott to children.
4. Lloyds Survey Register, No. 9353, National Maritime Museum, England.

Gold Digger

In a small isolated community such as that at the Swan River, the erratic state of communications was such that false rumours about shipping could go uncorrected for some time. During 1853, at a time when many people were leaving Western Australia to try their hand on the goldfields, it was speculated that two vessels, the *Gold Digger* and the *Gold Seeker*, had been lost.[1]

The cutter *Gold Seeker* sailed for Melbourne via King George Sound on 25 September, and on 5 October, the *Melbourne Argus* wrote:

> the cutter *Gold Seeker* left Swan River about three months ago, bound for Melbourne, with 30 or 40 passengers. Much anxiety was felt, no account of her arrival at Melbourne having reached the Swan.[2]

The *Inquirer* of 28 December reported:

> The *Gold Seeker*, it appears, was not wrecked at King's Island, having arrived safely at Melbourne on 30th October. No news has been received of the *Gold Digger*.[3]

Jack Loney, in his *Australian Shipwrecks*, writes that a schooner named *Gold Seeker* was wrecked on the Richmond Bar in 1861, and that a ketch named *Gold Digger* foundered between Newcastle and Sydney in 1862.[4] It may be that the cutter *Gold Digger* survived the voyage from Fremantle and later changed her rig.

NOTES

1. *Perth Gazette*, 21 October 1853.
2. *Melbourne Argus*, 5 October 1853.
3. *Inquirer*, 28 December 1853.
4. Jack Loney, *Australian Shipwrecks*, Volume 2 (Sydney, 1980), pp. 121, 131. The 1861 *Mercantile Navy List* has a 22-ton Melbourne-registered *Gold Digger* (Official Number 32665).

Leander

Local residents were saddened when the brigantine *Leander* sailed from Fremantle in November 1853. The master planned to proceed via northern ports and Surabaya to England, and not return to the Colony. Since 1851, the *Leander* had been very useful in keeping the Perth markets well supplied with produce from India and the East Indian Straits settlements. It was also one of the first substantial vessels (173 tons) to bother to venture into the smaller outports. The *Inquirer* reported:

> As the vessel which first settled the question as to the capabilities of Port Gregory as a harbour for ships of superior tonnage, and demonstrated that it was not a mere boat harbour, her name will be favourably associated with our Colonial History.[1]

Port Gregory was an outlet for lead ore, a convenient paying ballast for the lightly laden wool ships departing for London. Wool was a bulky but light cargo, so wool ships required some weight to balance the load.

Captain Johnston responded to the colonial hospitality by firing an early morning salute of twenty guns as he was departing, thus giving rise to the belief in Perth that a new Governor had arrived in the Colony.

As the *Leander* crawled northwards at 6 kilometres per hour, a death occurred among the East Indian crew:

> About 2 o'clock p.m. the Sarang came aft to the captain to say that 'Lascar die'. We then found that one of the men, whilst eating dinner, fell backwards, dead. He had been ailing a few days, and for two days had been off duty; he did not seem to suffer much, and until the moment of his death had eaten heartily. He was immediately brought on deck, and found to be quite dead. Preparations were made for his burial. The service was performed, in Mahomedan [*sic*] fashion, by the Bandare and subsequently by Captain Johnston, and at 5 p.m. he was buried.[2]

The next day, 13 November, as the *Leander* moved along the coast, a strong breeze increased to a half gale. At noon, Captain Johnston thought the vessel was in latitude 29° 30′, and at 4 p.m., as the wind increased, the vessel was hove to under close-reefed fore-topsail and mainsail. The intention was to drift into an open bay, ready to make sail and run into Champion Bay at daylight the next morning. Johnston perused his chart, which showed no reefs in the area, and went to bed.

However, his chart was of course incomplete and, in addition, he had misjudged his position. A few minutes before 10 p.m., a sailor rushed into the cabin to say he had seen broken water right ahead, and the brigantine hit the reef almost immediately. The vessel was carried over the reefs, resulting in the loss of her rudder, and she started leaking badly. The land could just be seen 16 kilometres away.[3]

Captain Johnston decided to anchor, but on hearing that there was 0.9–1.2 metres of water in the hold, he changed his mind and made more sail to drive the vessel ashore. The *Leander* was by then settling quickly, so the boats were launched and the crew waited astern until she struck the beach. Then they went back on board and took ashore a kedge-anchor and line, attaching one end to the ship to help the passage of the boats to and fro. Sails and spars were landed for making tents, but no water was saved, the casks all having been stove.

On the afternoon of the 15th, three men made an excursion inland with a horse saved from the wreck, hoping to find a road. They were turned back by impenetrable bush. Attempts by some of the crew to walk to Champion Bay were unsuccessful. Others set out along the beach for Fremantle, against the wishes of Captain Johnston, who considered their actions to constitute a mutiny. Yet these men had stayed to assist for more than a week after the *Leander* was wrecked, and left only because Captain Johnston took their rice away from them.[4] Johnston was apparently confused about his position. He thought the vessel was 6.4 kilometres north of the Irwin River, but later reports stated that it lay some 24 kilometres south of the river.

Eventually, on the 25th, one of the passengers, guided by a local Aborigine, reached a pastoral station on the Irwin and help was summoned. The Resident Magistrate, William Burges, surveyed the vessel and declared it a wreck.[5] By this time, a sand-bar had developed from the shore to the wreck and it could be reached by wading out, knee deep. Much of the cargo was saved and sold.

The turbulent Captain Henry Sanford, of the 43rd Regiment, lost some goods. He wrote to his brother, the Colonial Secretary, prefacing his news about the lost goods with some alarming comments about his own behaviour:

> I am sorry to say I have had a great row with C [Augustus Charles] Gregory, flung a tomahawk at his head, and cut his wrist severely. He angered me to madness, calling me a chisel and trying to cheat Govt out of some blankets . . . I am here [at the Irwin] for the last two days saving what I can from the *Leander* but shall be loser £40 to £50 pounds. G.M.C. [Geraldine Mining Company, the lead mine at Port Gregory] heavy losers, but fortunately not many had goods on board.[6]

Surprisingly, the wreck of the *Leander* has never been reported as found. It must be buried in sand very close to shore. On modern charts Leander Point is marked only several kilometres south of the mouth of the Irwin. When the ship came ashore, it struck adjacent to a large sand ridge that ran several kilometres inland and was a prominent feature. This description may suggest that the wreck lies close to White Point, further south of the Irwin. (See map on p. 44.)

The 173-ton brigantine *Leander* was built at New Brunswick in 1849 of birch, pine and spruce, and was fastened with iron bolts.[7] Her owners were Smith and Co., and she was registered at the Port of London with an A1 classification. The vessel left Fremantle with a light cargo, her draft having been 4 metres on arrival and 2.7 metres on departure. She was manned by twenty-one crew, and carried five passengers.[8]

NOTES

1. *Inquirer*, 16 November 1853.
2. *Inquirer*, 7 December 1853. Letter from Kenneth Brown, a passenger. A serang is the boatswain of an East Indian crew. A bandore is a three-stringed lute.
3. *Perth Gazette*, 9 December 1853.
4. Statement of E. Bruce (apprentice on *Leander*), 26 February 1854, C.S.R. 309, fol. 90.
5. William Burges (Resident Magistrate) to Col. Sec., 3 December 1853, C.S.R. 256, fol. 202.
6. Henry Sanford to William Sanford, 3 December 1853, Sanford Papers, National Library MS 818/3(6). A. C. Gregory, explorer and surveyer, carried out important work in the Geraldton area. He was later made Surveyor General of Queensland, and was knighted.
7. *Lloyds Shipping Register*, 1853.
8. Register of Arrivals at Fremantle.

Charles Fox Bennett

An outbreak of scurvy on board the *Charles Fox Bennett* caused the master to divert from his course across the southern Indian Ocean. The vessel was bound from Liverpool for Melbourne, but Captain Burns saw Augusta marked on his chart and stood in for the tiny port to get refreshments.[1]

At 1 p.m. on the morning of 18 November 1853, the *Charles Fox Bennett* was doing 2 knots before a light wind when she struck a sunken rock about 4 kilometres from Cape Leeuwin. Two hours later, the rudder came adrift. The crew set about lightening the vessel in the hope of getting her off the reef, but as night approached, the weather deteriorated, so the crew deserted and went ashore.

During the night, the unattended vessel drifted ashore, the falling tide the next morning leaving her resting on a rock ledge. But the following night, a heavy sea carried her off the rocks into a hole about 20 metres from the shore, where she filled with water, but lay relatively undamaged.

The crew worked industriously, and by the 28th had landed about 100 tons of cargo, including a steam sawmill, mounted on wheels like a locomotive to facilitate its hauling by a pair of horses.[2] The vessel was so close to shore that a landing stage could be laid down onto the beach for cargo to be taken off. There was no chance of saving the ship itself, because 10,000 bricks and some parts of 'iron houses' remained in the hold under 2.4–2.7 metres of water.

A sale of goods saved from the wreck realized the exceedingly high figure of £2500. The steam sawmill was bought by Henry Yelverton, the pioneer of the Western Australian timber industry, for £210.[3] Yelverton obtained leases and established a timber mill in 1853.

In 1980, skindivers David Biltoft and T. Eggleston showed Museum divers a wreck they had found in Ringbolt Bay, approximately 2 kilometres east of the Cape Leeuwin Lighthouse. They were given a reward of $250. The site, which was gazetted under the Historic Shipwrecks Act, lies at latitude 34° 21.3′ south, longitude 115° 9.2′ east. The wreck mound contained an estimated thirty to forty barrels of cement and overlay some timber structure. This may be the site of the *Charles Fox Bennett*, but further work will be necessary for a positive identification. It is possible that the material is from a concrete barge lost during the construction of the Cape Leeuwin Lighthouse.

Little is known about the *Charles Fox Bennett*, beyond the fact that she was a vessel of 127 tons and carried, besides a general cargo, a crew of ten men and fifteen passengers, including three women and a child.[4] The name does not appear on the Liverpool Register of British Ships.[5]

NOTES

1. *Perth Gazette*, 2 December 1853.
2. *Perth Gazette*, 16 December 1853.
3. *Perth Gazette*, 27 January 1854.
4. *Inquirer*, 30 November 1853
5. Personal communication from R. L. Dawson (Assistant to the Registrar of British Ships, Liverpool) to G. Henderson.

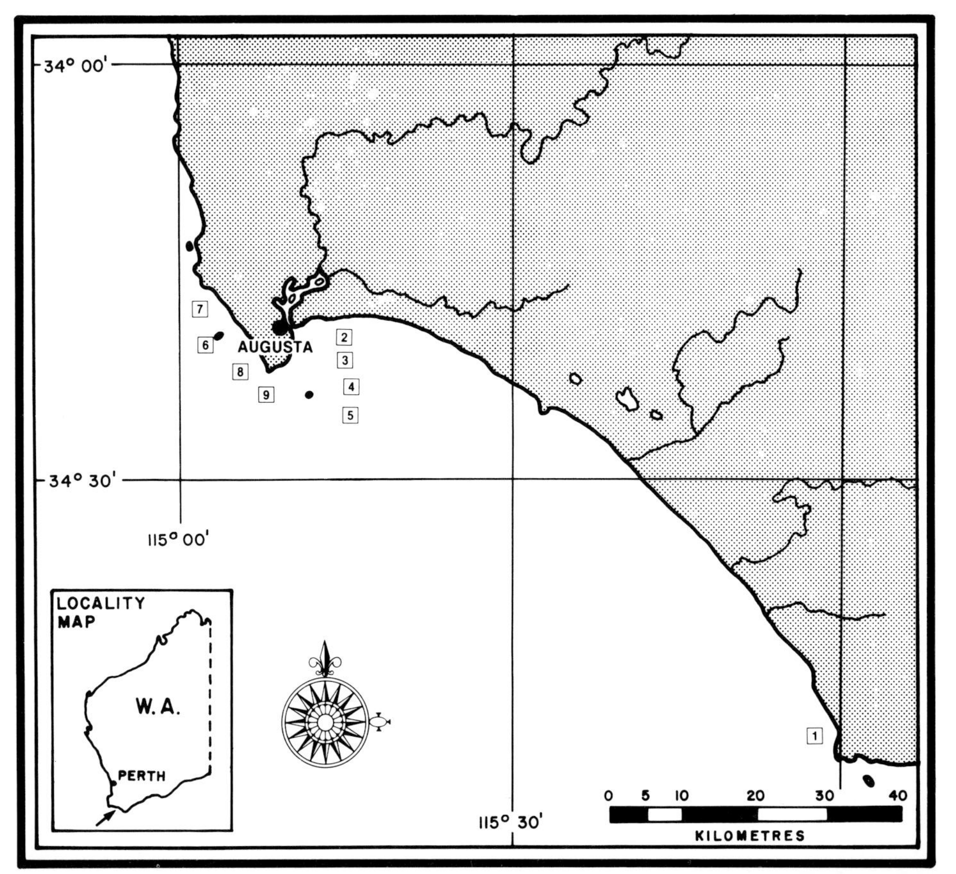

1 Whaleboat 1831
2 Boat 1839
3 *Salve*
4 *Charles Fox Bennett*
5 *MAYFLOWER*
6 *CUMBERLAND*
7 *Hokitika*
8 *Kitty Coburn*
9 *Alpha*

Northumbria

The brig *Northumbria* touched bottom in Gage Roads on 7 December 1853 while in charge of the Harbour Master, and unshipped her rudder, splintered part of her false keel and lost some copper sheathing.[1] The Colony's diving apparatus was used to examine the damage, which was not sufficient to prevent her sailing for Victoria in January 1854.

NOTE

1. *Perth Gazette*, 16 December 1853.

Unidentified Schooner, 1854

In January 1854, the crew of the cutter *Henry and Mary* reported that, while returning to Fremantle from Port Gregory, they saw the wreck of a schooner of about 60 tons burthen at a position 20 kilometres to the northward of the scene of the *Leander*'s disaster. The masts were still standing some 4.5 metres above water, but her hull was not visible. It was said that she had earlier been seen to pass Champion Bay in an abandoned state. (See map on p. 44.)

The *Perth Gazette* regretted that the authorities had no Colonial vessel available with which to investigate such reports.[1]

It seems unlikely that the *Venus* could have floated off the reef at the Abrolhos, or that the *Henry and Mary*'s crew would have confused this site with the wreck of the brigantine *Leander*. Yet the Resident Magistrate at Champion Bay appears to have made no mention of such a wreck in his correspondence during the first half of 1854.[2]

NOTES

1. *Inquirer*, 1 February 1854, Supplement.
2. Resident Magistrate, Champion Bay, C.S.R. 309.

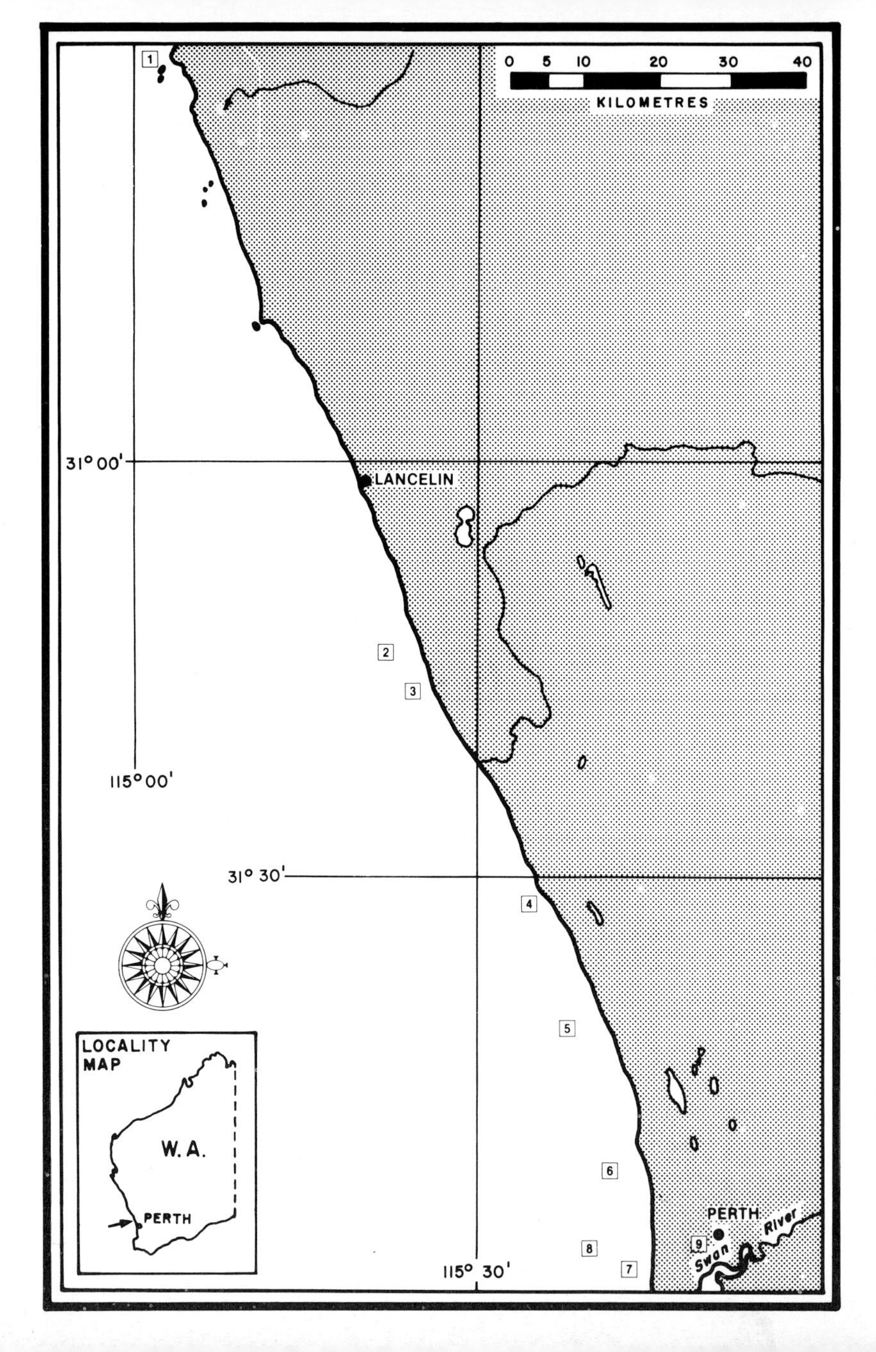

1 *CERVANTES*
2 *VERGULDE DRAECK*
3 *Sea Bird*
4 *Emily*
5 *EGLINTON*
6 *CENTAUR*
7 *ELIZABETH*
8 *Vixen*
9 *Canton*

Canton

In 1880, Thomas Browne, a civil engineer, took advantage of an exceptionally low tide in the Swan River to salvage portions of the wreck of the *Canton*, which had been an obstruction to river traffic. The wreck lay at the entrance to Miller's Pool at Mill Point, the present location of the Narrows Bridge. Browne raised pieces of sawn jarrah from the *Canton*, a river barge that had been wrecked in 1854.

He used the sawn jarrah boards as evidence to support his argument that squared wooden timbers were no more subject to rot or other deterioration than were those left round, when used in construction.[1]

The *Canton* was described as having been a large river barge or flat, built of jarrah and owned by Mr Habgood. Peter Habgood was a merchant and shipping agent.

NOTE

1. 'Round or square'. Letter to the Editor of the *Herald*, 30 April 1881.

A barge on the Swan River, around 1870.

Mary Queen of of Scots

The barque *Mary Queen of Scots* left Fremantle bound for Singapore via Port Gregory on 5 February 1855. She sailed with a complement of forty-seven, made up of thirteen crew, five steerage passengers, two ostlers (stablemen), twelve ticket-of-leave men, and fifteen prisoners. The cargo consisted of twelve horses and some goods for the Commissariat and the Geraldine Mine at Port Gregory, where 40 tonnes of lead ore was to be taken on for Singapore.[1]

On the afternoon of the 7th the vessel sailed in through the Hero Passage at Port Gregory under a south-south-west wind and anchored inside the reef. Shortly after, she began to drag, so another anchor was dropped. But at about 10 o'clock in the evening, a gale developed and she dragged and began striking heavily about 180 metres from shore. By 3 a.m., her back had been broken, and an hour later there was 2.4 metres of water in the hold. The horses were hoisted on deck and put overboard to swim ashore. (See map on p. 53.)

Eventually, a whale boat was able to bring a line to the barque from the shore and three men used the ship's boats to reach the land.[2]

On the 12th, the weather settled enough for some men to go back on board and save part of the cargo. The next day, the mainmast parted, carrying with it so much of the rigging that it was impossible to get near the wreck. The vessel then went to pieces and disappeared entirely below the water.[3]

The hull, anchors etc. were sold by auction and realised a mere £80. The four boats were sold for £20.[4] The owner, Thomas Sleddon, who had been on board when the vessel struck, lost heavily as the vessel was not fully insured. The question of blame occupied the minds of local residents, who feared that the Government would abandon the Port in the wake of the disaster.[5] The Resident Magistrate laid the blame with Captain Buxey for having anchored overnight in an unsafe part of the anchorage 'just at the tail of the reef on the southern side of the Hero Passage where she felt the influence of the heavy current both from this and the Gold Digger Passage'.[6]

The 256-ton *Mary Queen of Scots* was built at Sunderland and registered at Liverpool.[7] Her dimensions were 28.7 metres by 6.6 metres by 4.7 metres. She had one deck and a break, and was barque rigged, square sterned and carvel built, with a standing bowsprit and a figurehead of a woman.

NOTES

1. *Inquirer*, 28 February 1855.
2. *Commercial News*, 1 March 1855.
3. Resident Magistrate, Champion Bay to Col. Sec., 13 March 1855, C.S.R. 339, fol. 6.
4. *Commercial News*, 29 March 1855.
5. *Inquirer*, 28 March 1855.
6. Resident Magistrate, Champion Bay, to Col. Sec., 13 March 1855, C.S.R. 339, fol. 6.
7. Register of Arrivals at Fremantle.

Stag

The Colony of Western Australia began receiving transported British convicts in 1850. On 24 May 1855, one of the convict transports, the *Stag*, arrived at Fremantle with 225 convicts and a large number of other persons on board, totalling in all approximately 500.[1] As the *Stag* was being brought in, under the charge of Pilot Goss, strong winds were blowing and the vessel grounded on Success Bank. Captain Clarke then 'took charge out of the Pilot's hands, he evidently not being a fit person to have charge of a ship containing so many lives'.[2]

The Harbour Master set off to render assistance and the Resident Magistrate at Fremantle, seeing that the *Stag* continued to signal for help, sent off several cargo boats to try to reach the vessel. The signals still continued, so George Clifton of the Water Police, Captain Bennett, late master of the *Eglinton*, and Pilot Back, also set out. Arriving on board, they found that the Harbour Master had turned back, and refused to come to her assistance, saying that no boat could live in such a sea.[3] The *Stag* was striking so heavily that the surgeon superintendent on board decided to land the prisoners and pensioners with their wives and families. All were safely into the boats when, at about 11 p.m., the ship floated off with the increasing tide and was conducted into Owens Anchorage by Pilot Back.[4] The *Stag* suffered no major damages and sailed for India on 28 June.

The 545-ton Sunderland-built ship *Stag* was commanded by Captain H. Clarke and had a crew of twenty-nine. Her draft on arrival was 4.8 metres.[5]

NOTES

1. F. Bray, 'George Clifton, Inspector of Water Police', *R.W.A.H.S.J.* 2: XX (1936), p. 12.
2. Henry Clarke, Log Book of the Convict Ship *Stag* from Deptford to Western Australia. Battye Library, 276A.
3. Bray, *op. cit.*, p. 12.
4. Resident Magistrate, Fremantle, to Col. Sec., 26 May 1855, C.S.R. 339, fol. 165.
5. Register of Arrivals at Fremantle.

Iris

The 311-ton New Bedford whaling ship *Iris* fished in Pacific waters in the 1820s and 1830s, and shifted to the Indian Ocean grounds in the 1840s and 1850s. In December 1851, she was reported at Geographe Bay with seventy barrels of 'black oil'.[1] In August 1854 the owner, E. C. Jones, sent her on another voyage to the Indian Ocean. When she had been out for ten months, the old whaler ran into some problems.

The *Iris* entered Port Gregory on 29 June 1855. She had sprung a leak, and her captain, Edward Devoll, wanted to find a sheltered spot to heel her over for repairs.[2] Eleven days later, the ship had been blown ashore and lay high and dry on the sand spit opposite the Gold Digger Passage. The Resident Magistrate, William Burges, bewildered by the event, wrote:

> It was during a heavy gale from N.W. and strange to say the current carried her out against the wind, and the anchors seemed to drift faster than the vessel as if the whole bottom of the anchorage bodily gave way. She had an anchor out astern and it, all the time she was dragging, kept a strain on the cable. She went out stern foremost and had three anchors

> out, one of which came foul of the mooring chains of the large buoy, and all went away together . . . It is a curious fact that the vessel is now lying with her three anchors and the mooring buoy and anchor astern of her.[3]

Captain Devoll and his crew (thirty all told, including men and boys) pitched tents on the shore and commenced unloading the vessel with the hope of refloating her. They made no progress, and in August, a board of survey condemned the vessel as a wreck. The crew took passage to Fremantle on the schooner *Perseverance* to claim assistance from the American Consul residing there, Thomas Pope. The issue of the responsibility for the maintenance of destitute sailors had been a problem for Burges. He understood the American Consul to be responsible for such men from American vessels. However, some of the crew were Portuguese and, as such , were not given assistance, although the Advocate General pointed out that the Consul's responsibilities 'do not exclude a regularly hired seaman on board an American vessel, but only foreign precarious adventurers of doubtful character'.[4]

In October, it was announced that goods saved from the wreck would be auctioned, including 'bread, slops, clothing, casks, trypots, copper cooler, sails, spars, rigging . . . superior Whaling Gear, comprising everything fit for a season'.[5] It seems that the hull, spars and rigging were bought together as one lot, because in January 1856, it was announced that the *Iris* was again afloat, that the leak had been located and stopped, and that she could be expected at Fremantle shortly. Captain Bennett (of the *Eglinton*) was one of the new owners, and he had engaged a working party to perform the difficult job of getting the *Iris* off the beach. At least one of his workers, the freeman William Johnson, had no stomach for the work and absconded. He was apprehended and sentenced to three months' imprisonment with hard labour![6] Bay whaling crews at Port Gregory and elsewhere registered with the local magistrate, and the remedy for breach of contract was set in 10 Vic. No. 16, an ordinance for contracts in the fisheries.

The *Iris* set out for Fremantle under an inadequate jury-rig, which made her so slow that, for a time, she was posted as missing.[7] The vessel was unable to make headway beyond Champion Bay and so returned to Port Gregory for further work on her rigging. Captain Bennett finally brought her into Fremantle on 24 April. She was then taken to Careening Bay to be hove down and repaired.[8]

Prior to 1850, it would not have been legally possible for a 'foreign'-registered vessel to be repaired after having been condemned as a wreck, because of the provisions of the British Navigation Laws.[9]

When the vessel left Careening Bay for Adelaide on 12 October 1856, she had been re-registered as a British vessel under the Merchant Shipping Act of 1854, given a unique number (Official Number 40474), renamed the *Frances*, and re-rigged as a barque. The *Frances* carried 150 emigrants to Adelaide, principally artisans who had found themselves displaced by convict labour in Western Australia.[10] Under the ownership of the importer Richard Clinch, the *Frances* made several voyages to London. Her last visit to Fremantle was in April 1860, after which she sailed to Madras with a cargo of timber. She was struck from the Register in 1882.[11]

The *Frances* was a 271-ton barque with three masts, two decks, a standing bowsprit, square stern, a scroll head and no galleries. Her dimensions were 29.6 metres by 8.4 metres by 5 metres.[12]

NOTES

1. *Inquirer*, 8 January 1851.
2. William Burges (Resident Magistrate, Champion Bay) to Col. Sec., 30 June 1855, C.S.R. 339, fol. 57.
3. Burges to Col. Sec., 13 July 1855, C.S.R. 339, fol. 67.
4. Advocate General to Col.Sec., 28 August 1855, C.S.R. 339, fol. 80.
5. *Inquirer*, 31 October 1855.
6. Burges to Col. Sec., 14 December 1855, C.S.R. 339, fol. 128.
7. *Inquirer*, 20 February 1856.
8. *Inquirer*, 30 April 1856.
9. See *Unfinished Voyages 1622–1850*, p. 203.
10. *Inquirer*, 15 October 1856.
11. Register of British Ships, Fremantle.
12. Register of Arrivals at Fremantle. Register at rear of volume.

Occator

The brigantine *Occator* sailed out of Melbourne on 12 January 1856 bound for North West Cape. The vessel had been chartered by J. F. Jones of Melbourne, who had been told that on Muiron Island, just off the Cape, the deposits of guano were 2.4 metres thick.[1] Jones joined the voyage as a passenger.

At 4 p.m. on 4 February, Captain Place estimated that he was 80 kilometres west of the Cape, so he tacked to the east in rough seas but light winds. Towards midnight, the wind freshened a little, and the mate relieved Place on the watch. The captain said that he expected to see land in the morning at 20–25 kilometres' distance. At about 3.30 a.m., however, the mate called out 'put the helm hard down', and called the captain.[2] Place came on deck at once, and seeing the vessel in stays, gave orders to haul the yards round. The order was executed promptly, but as the vessel was paying off, her heel caught the rocks, and the surf was running so high that a heavy sea broke over the vessel and threw her right onto the reef. Shortly after, the rudder was unshipped, and the ballast was found to be covered with the rising water.

The first boat to be launched was knocked to pieces by the breakers. Then the crew, with much difficulty, launched the longboat. The mate went ashore with Jones and four hands, while the captain brought up some provisions from the lazarette. On the boat's return, the captain and the remainder of the crew jumped in, taking with them 164 litres of water, several bags of bread, some spirits and preserved meats, and some of the ship's navigational equipment.

The ten men sailed 90 kilometres north to Muiron Island, where they hoped to meet up with another vessel (the *Eblana*). After approaching the island through a narrow passage in the reef, they were disappointed to find no sign of the other vessel. Four days later, their tongues parched and blackened from thirst, they decided to return to the wreck for more water. Seeing that it was impossible to relaunch the longboat through the passage where they had beached her, the crew were obliged to carry the craft 3 kilometres across the island to a more sheltered launching spot. Their troubles were far from over:

> In sailing down from the North West Cape a party of about 40 natives appeared on shore, and followed the boat as far as the wreck. Their demeanor did not appear unfriendly till the boat, being some distance in advance of them, ran in for a cask of water that had been washed ashore from the wreck. Before the natives came up the boat had got out a safe

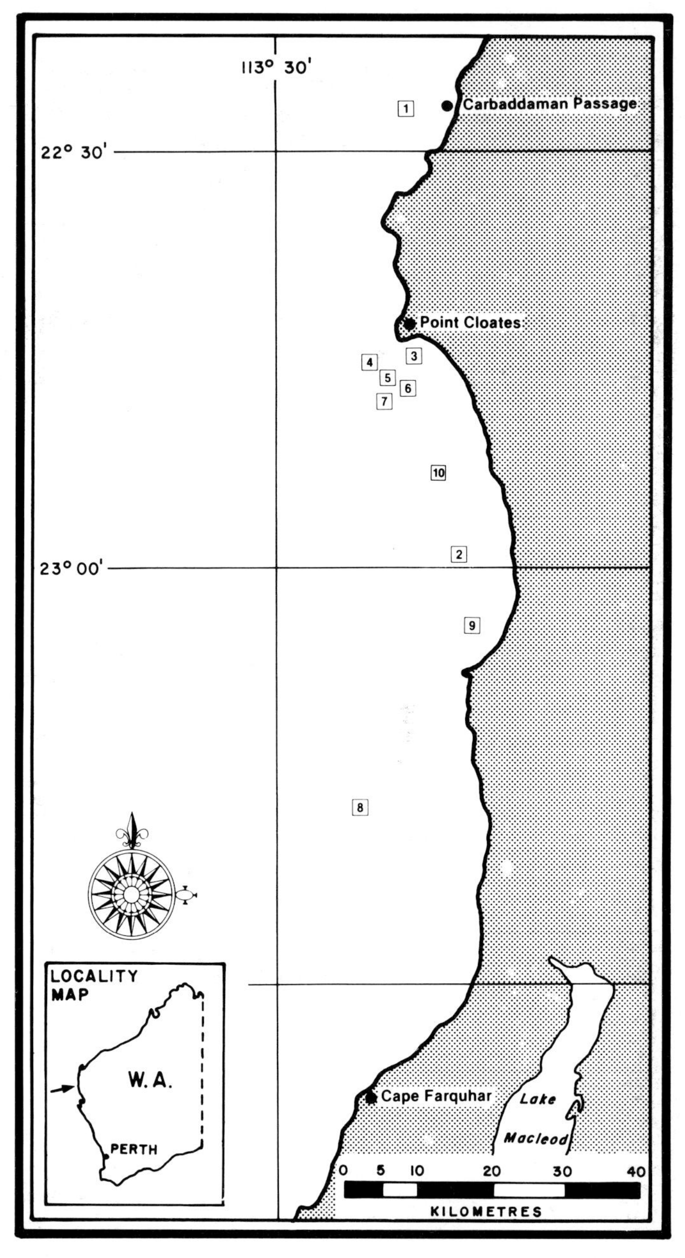

1 *Occator*
2 *Correo d'Azia*
3 *RAPID*
4 *Strathmore*
5 *Stefano*
6 *Bertha*
7 *Cock of the North*
8 *Brothers*
9 *Emma*
10 *Caledonia*

> distance, and now, as if suspicious that something useful had been carried away from them, they put themselves into menacing attitudes, and next walked into the water after the boat, armed with stones. When the water became too deep for wading, about 10 of the more hardy swam out with the evident intention of seizing the boat. When they got nearer, one pitched a spear at the boat, but this was caught by one of the men without doing any harm, and others cast stones with slings. Mr Jones at this time fired, and shot one of them with a charge of slug and shot. They then retired and lighted a fire on the hills.[3]

From the wreck, the party proceeded to Shark Bay. The voyage there took 4 days, but once in the Bay, they spent another 4 days beating about, trying to reach Dirk Hartog Island. When they finally got ashore, they were so weak that they could hardly stand.

The men lived on the island for about a month, sucking the blood of turtles as a substitute for water, and catching crabs and a few wallabies. They made several vain attempts to proceed in their boat to Fremantle, and then crossed to what they thought was the mainland, with the intention of walking to Fremantle. But they found to their dismay that they were on another island. The men were returning to Dirk Hartog Island when they saw the schooner *Favourite* lying at anchor off Bird Island, where her crew were collecting guano. It had been 42 days since they lost their ship.

The wreck of the *Occator* has not been found in modern times. Contemporary reports indicate that the vessel struck between 48 and 64 kilometres south of the Cape, which would place it in the region of Carbaddaman Passage.

The 145-ton brigantine *Occator* was built at Prince Edward Island in 1853, but was owned in Liverpool by G. and R. Barclay.[4] Her dimensions were 27.2 metres by 6.2 metres by 3.5 metres. She was copper and iron fastened, with ten pairs of iron hanging knees, and had been sheathed with yellow metal.

NOTES

1. *Perth Gazette*, 11 April 1856.
2. *South Australian Register*, 12 May 1856.
3. *Ibid*.
4. Lloyds Survey Register, National Maritime Museum, England.

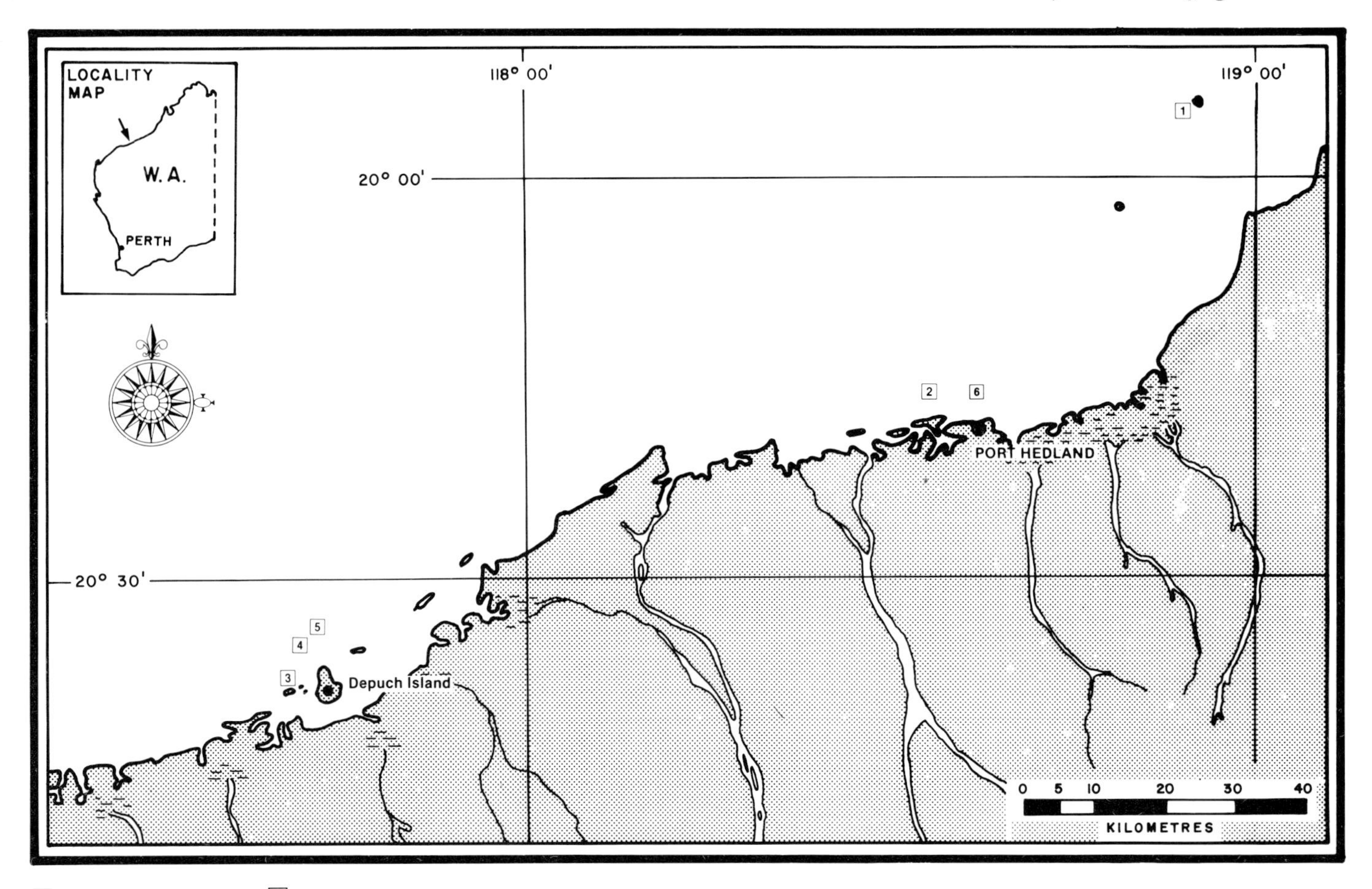

1 *North Star*
2 *Sea Spray*
3 *Melanie*
4 *Young Victorian*
5 *Chip*
6 *Pearling Boat, 1872*

North Star

The whale ship *North Star* was built at Philadelphia in 1828. In 1850, the vessel was sold to Williams and Barnes of New London, Connecticut, and sent on a whaling voyage.[1] Her recorded destination on this 5-year voyage was the North Pacific, but the vessel ranged more widely in search of whales. In February 1851, she put in at Geographe Bay for supplies. There the enterprising and inventive Captain Robert Brown sold £500 worth of harpoon guns to the other American ships in harbour. Brown was the inventor of the gun, which was made almost entirely of gun metal, weighed a mere 16 kilograms, and was fired from the shoulder by the harpooner standing in a whale boat.[2] The correspondent for the *Inquirer* was impressed:

> It has this advantage over the ordinary harpoon, that the harpooner can make fast to a whale at the distance of 18 fathoms; with the weapon usually employed, it is necessary to get pretty close to the whale, and there is consequently more risk of frightening them, in which case it is ten to one against the boat steerer having another opportunity of attempting a stroke.[3]

The harpoon gun did not come into the whaling industry suddenly. One was invented in 1731, and Abraham Staghold produced an improved version in 1772.[4] In the early years of the nineteenth century, there were several attempts to introduce harpoon guns. In 1821, Scoresby's ship *Baffin* made a voyage to Greenland with a new harpoon gun aboard.[5] But the harpooners, like English bowmen of the fifteenth century, hated the smell of gunpowder. The importance of Captain Brown's invention was that it was the first, although perhaps the heaviest, of the shoulder guns.

It was in the context of a dying industry that the most significant invention was made. In about 1864, the Norwegian sailor Svend Foyn designed a gun and harpoon on which all later weapons have been based. The gun was a steel muzzle-loader, about 1.2 metres long, and mounted on swivels; the projectile a combination of harpoon and bomb. This invention was far-reaching in its implications because it not only greatly enhanced the chances of hitting whales up to 45 metres away, but also increased the likelihood of remaining fast to them once they were struck. It also made possible the pursuit and capture of the fin whales, which had hitherto been beyond man's reach because they sank promptly when dead. The device did away once and for all with the whale boat, for a 1-ton gun had to be mounted on a large vessel. This snag held up the adoption of the gun for 15 years, as there were plenty of vessels big enough to carry the gun, but none with sufficient speed and manoeuvrability to shoot the fin whales. Introduction of the harpoon gun had to await the development of another Norwegian invention, the whale chaser.[6]

Captain Brown's catch, using his hand-held gun, was quite good. During his early fifties voyage, he sent home 2,007 barrels of whale oil and 10,200 kilograms of whale bone, and when he returned to New London in June 1855, he had on board another 660 barrels of oil and 3,720 kilograms of bone.[7]

Williams and Barnes lost no time in turning the *North Star* out of port for another similar voyage. She left for the Pacific Ocean on 11 September 1855 under the command of Silas Fish. Julia Fish travelled on board with her husband. On 4 May 1856, with a crew of thirty-six men, the *North Star* entered Fremantle, ostensibly to recruit more crew. Mrs Fish went ashore for a time, but was not impressed with Fremantle. She wrote to her sister:

> I think I enjoy myself as well on board the ship as I should to stay in this place, for I try to think we have seen the roughest part of the voige [*sic*]. I have seen times when I had rather been on land but if I ever have the good fortune I shall never be sorry I went to see [*sic*] for I shall see a great many different scenes, I can say now I never new [*sic*] what the world was composed of before. When I see you if we are spared to meet, I can tell you some things that will make you laugh and wonder too. There was a ball at the Hotel where we stop so I just went in to see them. They had a brass band to play for them. There was married women there with nursing children, Irish girls, soldiers etc and such an [*sic*] brandy and gining time I never witnessed. Ladies and all will drink. Not much aristocracy here.[8]

The *North Star* and her crew of thirty-seven left Fremantle for the whaling ground on 28 May, and ran into trouble 6 weeks later. At 8 p.m. on 11 July the

North Star was sailing north-easterly towards Bedout Island, 24 kilometres away.[9] A brisk gale sprang up from the south-east and the ship headed between east and east-north-east. At 1.40 a.m. the following morning, the man in the chains called 13 metres. The crew tried to drop the anchor, but the line tangled and the vessel drifted onto a small coral reef near Turtle Island. Captain Fish wrote:

> Hauled the yards aback, and the ship drifted off and wore partly round and struck again, took in all sail and furled it. Before daylight the rudder broke away . . . breaking every pintle, breaking up the stern, and after part of the deck, the sea breaking clear over the ship . . . At 7 o'clock a.m., sent a boat with the chief officer to seek help from the *Vesper*; all hands were then set to work to lighten the ship in hopes that by a full sea she might be got clear. At 3 o'clock p.m., the sea having gone down, an anchor was got out to haul off the ship . . . although she apparently was broken in two, her upper deck being raised nearly a foot amidships.[10]

The crew went ashore on Bedout Island where, 2 days later, Captain Hempstead of the American whaler *Vesper* arrived to provide assistance.

There was no hope of saving the ship, but 835 barrels of oil, a quantity of whale bone and many other articles were salvaged. On 16 July, the wreck and oil were sold to Captain Hempstead. The *Vesper* sailed for Port Louis on 1 August with the entire crew from the *North Star*.

Since that time, several groups have seen wreckage in the area. In 1863, the surveyors Charles Hunt and Joseph Ridley landed at the De Grey River and explored the surrounding country.[11] They reported finding wreckage at North Turtle Island.[12] In 1867, parts of the frame of a vessel were found about 10 metres above high-water mark on Bedout Island.[13] Iron hoops (from oil barrels) and 'children's shoes' (perhaps belonging to Mrs Fish), were also found on the island.[14]

The 399-ton *North Star* was ship rigged with two decks, three masts, a square stern and a billet head. Her dimensions were 33.5 metres by 8.5 metres by 3.3 metres.[15]

NOTES

1. New London Registry. Copy sent to G. Henderson by Lt. Col. (U.S.A.F. Ret.) James Keesling.
2. *Inquirer*, 5 February 1851, Supplement.
3. *Ibid*.
4. E. F. Carter (ed.), *Dictionary of Inventions* (New York, 1976).
5. F. Morley and J. Hodgson, *Whaling North and South* (London, 1927), p. 64.
6. Ivan Sanderson, *Follow the Whale* (London, 1956), p. 299.
7. A. Starbuck, *History of the American Whale Fishery* (New York, 1964), p. 474.
8. Julia Fish to Clara Burrows, 14 May 1856. Family Records of Lt. Col. James Keesling.
9. *Whaleman's Shipping List and Merchant's Transcript*, XIV: No. 38, November 1856. Copy sent to G. Henderson by Lt. Col. James Keesling.
10. *Ibid*.
11. J. S. Battye, *The History of the North West of Australia* (Perth, 1915), p. 68.
12. *Underwater Explorers Club News*, June 1967.
13. *Inquirer*, 11 September 1867.
14. Letter to the Editor, *Inquirer*, 18 September 1867.
15. New London Registry.

Sara

The 54-ton schooner *Sara* was built at Preston Point on the Swan River in 1855 by William Owston, master mariner, shipbuilder and importer. Owston succeeded in floating the *Sara* over the bar at Fremantle into the ocean, where she was prepared for a voyage to Adelaide.[1] But the vessel was to have a short life.

On her return from Adelaide, Owston sent her to Port Gregory under the command of Henry Christians for a cargo of oil and copper ore.[2] Bay whalers had been operating at Port Gregory since 1853 and were an important part of the regional economy. In 1856, they produced £1,600 worth of black oil and 1.27 tonnes of bone, compared with only £1,200 worth of lead ore from Northampton.[3]

The cargo was loaded and the *Sara* was being towed out of Port Gregory by three whale boats on 17 July 1856 when a heavy swell sent her ashore on a shoal about half-way between the Gold Digger Passage and the Hero Passage. All hands were saved, but the *Sara* went over on her beam ends and half filled with water.[4] In December 1856, the *Inquirer* reported that all efforts to get her off had been unsuccessful.[5] (See map on p. 53.)

The *Sara* (Official Number 31549) was 16.9 metres by 5.6 metres by 2.7 metres. She carried two masts and a standing bowsprit, and was of carvel build, with a square stern, no galleries and a scroll head.[6]

NOTES

1. *Inquirer*, 17 October 1855.
2. Register of Arrivals at Fremantle.
3. I. D. Heppingstone, Whaling at Port Gregory, PR7666, Battye Library.
4. Resident Magistrate, Champion Bay, to Col. Sec., 21 July 1856, C.S.R. 367, fol. 85.
5. *Inquirer*, 10 December 1856.
6. Certificate of British Registry, HS100, Battye Library.

Antelope

Bringing sailing vessels, particularly those with full cargoes, into the mouths of small rivers is generally a tricky business. At the mouth of the Swan, lighters had to contend with a sand-bar, rocks, and variable winds and currents.

On 27 August 1856, the cargo boat *Antelope* was laden with goods from the *Empress* consigned to Messrs W. and R. Habgood when the man in charge attempted to negotiate her into the Swan River. He had been advised that the current was running out too fast for the boat to get over with the light wind. However, he persisted, and the vessel was wrecked on the bar. The *Antelope* sank between two rocks, and soon went to pieces.[1] (See map on p. 242.)

NOTE

1. *Inquirer*, 27 August 1856. See also *Unfinished Voyages 1622–1850*, p. 215.

Inbat

The river cargo boat *Inbat* suffered a similar fate to the *Antelope*. She was proceeding from the Swan River to the schooner *Perseverance* in Gage Roads in December 1856 when she struck the bar across the river mouth. The vessel was towed into the South Bay in a sinking state. The goods on the boat, which were partially damaged, belonged to the pioneer merchant George Shenton and were not insured.[1] (See map on p. 242.)

The *Inbat* was a 5-ton sailing boat owned by John Watson, who specialized in boat services on the river.[2]

NOTES

1. Receiver to Col. Sec., 5 December 1856, C.S.R. 369, fol. 80.
2. *Inquirer*, 30 November 1853.

Unidentified Wreck, King George Sound, 1857

In January 1857, portions of the wreck and cargo of a ship were found near King George Sound.[1] The *Perth Gazette* referred to wreckage strewn along the coast to the west of Albany and gave a possible explanation:

> On October 31, latitude 45° 39′ and longitude 104°, being at the time under close reefed canvas, the *Albion* sighted within a distance of twenty yards, the wreck of a large ship, reported by Captain Angel to be fully 1,000 tons. Her mainmast only was standing, and on the stump of the maintopmast was lashed a topgallantmast, with signal halyards rove, for the purpose of hoisting signals of distress. The hull, although totally submerged, was visible from aloft apparently 40 feet below the surface.
>
> On reference to the map, it will be found that the locality indicated is about midway between Cape Lewin [*sic*] and Kerguelan's Island, and not much less than one thousand miles distant from the shore, certainly a most extraordinary locality for a shoal to exist in. One circumstance certainly against the wreck on the south coast being derived from the sunken ship seen by the *Albion* is, that as her mainmast was then standing, she must have sustained no great amount of damage in her hull, and the cargo

> consequently could not have got adrift, but she might have broken up during the two months which had passed between then and when the articles were found. It is possible the original cause of wreck was having the bows stove in from striking on an iceberg.[2]

In March, the *Inquirer* contributed to the mystery, reporting that:

> pieces of wreck continue to be picked up along the coast in the neighbourhood of King Georges Sound, and principally consisting of candles and boots and shoes. The figurehead of a ship has also been seen. We hope that next mail will bring intelligence that some fragment has been found which will lead to the discovery of the name of the wrecked vessel.[3]

But nothing further was reported. (See map on p. 126.)

NOTES

1. *Western Australian Almanac*, 1883.
2. *Perth Gazette*, 30 January 1857.
3. *Inquirer*, 18 March 1857.

Champion

The brig *Champion* was built at Great Yarmouth in the county of Norfolk, U.K., in 1839. She was later brought to Australia and, in 1853, was registered in Adelaide.[1]

The vessel made her first visit to Fremantle in the winter of 1856. On that occasion, the master moored her in Gage Roads for the convenience of being close to the port. But a storm blew her ashore, and it was estimated that it would cost several hundred pounds to get her afloat.[2]

The *Champion* was refloated, but soon afterwards got into more serious trouble at Busselton. She had almost completed loading a cargo of timber for Adelaide when she was blown ashore on 14 May 1857, in company with the coaster *Sea Gull* and a boat belonging to Henry Yelverton. She was pulled off the shore only to go back again during the next gale. On this occasion, she was condemned as a wreck and sold by auction.[3]

The enterprising Yelverton bought the wreck with the intention of repairing and re-registering her for use in the timber trade. After refloating the *Champion*, he took her to Careening Bay for repairs and then employed her in trade with Adelaide. Yelverton sold the vessel in 1861 to Hong Kong shipowners.[4]

The *Champion* was a 225-ton brig with the dimensions 27.3 metres by 6.7 metres by 4.4 metres. She was carvel built, with a standing bowsprit, round stern, no galleries and a male figurehead. The *Champion* was originally given Official Number 36536, but Yelverton re-registered her as 40486.[5]

NOTES

1. Register of British Ships, Port Adelaide.
2. *Inquirer*, 3 September 1856.
3. Register of British Ships, Port Adelaide.
4. Register of British Ships, Fremantle.
5. *Mercantile Navy List*, 1861.

Enterprise

A log book of the New Bedford ship *Congress*, under the command of Captain Hamblin, describes a visit to Flinders Bay, to the east of Cape Leeuwin, on 17 June 1857. The captain went up the river shooting and bagged ten black swans. Later, he found the wreck of a Liverpool ship, which he stated to be the *Enterprise*.[1]

Many vessels bore that name, although Lloyds does not list any such Liverpool-registered vessels for the early 1850s. It seems likely that Captain Hamblin saw the *Charles Fox Bennett*, which had come from Liverpool, and got the name wrong. Alternatively, the vessel may have had its name changed from *Enterprise* to *Charles Fox Bennett*, and Captain Hamblin saw the old name.

NOTE

1. W. J. Dakin, *Whalemen Adventurers* (Sydney, 1963), p.119.

Robertina

On 2 November 1859, the brig *Robertina* left Fremantle under the command of Captain Frederick Davis, carrying a cargo of timber, flour and whale oil intended for the Adelaide market. The vessel was piloted out by the Harbour Master, and then Captain Davis took a westerly course until 3 p.m., when he tacked the brig back towards the land on a south-easterly course to make use of the evening land breeze.

At 6 p.m., Chief Officer Joseph Mallison took cross-bearings. He could see Garden Island to the port side and the loom of the mainland ahead. The Coventry Reef lay about 4.8 kilometres off the lee beam. Captain Davis had a general chart of the area on board, but it did not show any reef in the vessel's path. The topgallant sails were taken in between 6 and 6.15 p.m. and orders were given to keep a sharp look-out.

At 6.50 p.m., Captain Davis gave the order 'all hands ready to bout ship', but ten minutes later she struck.[1] The sea was smooth, the weather was fine, and no breakers were visible. The lead was hove, showing 3.3 metres of water. The *Robertina*'s draft was 3.3 metres.[2]

The *Robertina* had struck the Murray Reef at a position less than 10 kilometres from the Coventry Reef. She went down immediately, head foremost, leaving about 1 metre of her stern above water.[3] The twelve crew and seven passengers had barely sufficient time to reach the boats before she sank.

An inquiry was held into the cause of the sinking of the *Robertina*, resulting in Captain Davis being charged with neglect of duty. However, the court found him not guilty. The wreck was sold for £30, and a fair portion of the cargo was sold at auction.[4]

Although the vessel's certificate of registry has the comment 'vessel wrecked in 1859 and registry closed',[5] the vessel was listed in the *Mercantile Navy List* as late as 1872.[6]

The wreck of the *Robertina* was re-located in 1987 by diver Graham Anderton. The site was identified by the bell, which had the inscription '*Robertina* 1843'. (See map on p. 242.)

The 213-ton *Robertina* (Official Number 31510) was built at Greenock in 1843, but by 1854 was registered at Melbourne. Her dimensions were 26.3

metres by 6.2 metres by 4.6 metres, and she had one and a quarter decks.[7] *The Robertina* had a square stern, and was carvel built, with a standing bowsprit and female bust figurehead. Her frame and planking were of wood.

NOTES

1. 'Quarter Sessions', *Inquirer*, 11 January 1860.
2. Register of Arrivals at Fremantle.
3. *Perth Gazette*, 4 November 1859.
4. *Inquirer*, 16 November 1859.
5. Register of British Ships, Melbourne.
6. *Mercantile Navy List*, 1872.
7. Register of British Ships, Melbourne.

Gazelle

A cargo boat named *Gazelle* was lost off Fremantle on 5 July 1860. The *Gazelle* was proceeding to a vessel named the *Rubens* when she disappeared in a heavy squall. The first indication of the accident to those on shore was the discovery the next day of the body of Mr Jones, the man in charge. Two weeks later, neither the wreck nor the body of the one passenger, a young gentleman named Julius Brockman, had been found.[1] (See map on p. 242.)

Several small Fremantle vessels bore the name *Gazelle*, but there is no indication that this one, owned by merchants J. and W. Bateman, was registered.

NOTE

1. *Inquirer*, 18 July 1860. Julius was the son of pastoralist William Locke Brockman.

Cochituate

The 347-ton American barque *Cochituate* was bound from Melbourne to Singapore under the command of Captain George Bangs when she struck the Abrolhos Islands at 3 a.m. on 14 June 1861.[1]

The vessel had experienced heavy gales for some days prior to the accident, but when it occurred, a light wind was blowing. The breakers were seen too late, and the helmsman had his shoulder bone broken by the spokes of the wheel at the time of impact.

Captain Bangs, the second mate and three crewmen left in one boat, while the first mate and six crewmen left in the other. The ship was half full of water and breaking up. Nothing was saved but what they were wearing. The two ship's boats sailed southward along the coast, but on the second day, rising winds forced them to go ashore and abandon their craft near the Arrowsmith River, a little south of Dongara.[2]

The second mate walked ahead with some of the crew, living on dead fish and birds that they picked up along the beach, and a dog they killed at the Moore River. At this point, they were getting very weak, so they gave one of their number, John Barlish, an extra allowance of dog and sent him on ahead. He reached Fremantle on 30 June, and the next day, a rescue party of mounted troopers reached the forward party, and the captain's party was found on 2 July. John Lee, one of the policemen, reported:

> . . . made the beach about 10 miles south of the Moore River. We again see the tracks and after following them about a mile came up with the Captain, Mate and one Man of the shipwrecked crew. They were in a dreadful state of starvation. I gave them a little brandy and water with a piece of bread, made a fire, boiled some sago, gave them a little which greatly refreshed them.[3]

The wreck was purchased by John Wellard, a survivor from the wreck of the brig *James Matthews* in 1841. He sent the cutters *Mystery* and *Speculator* to the Abrolhos to recover whatever remained. The headboard of the *Cochituate*, with her name in full, was found on Pelsart Island. The cutters then pro-

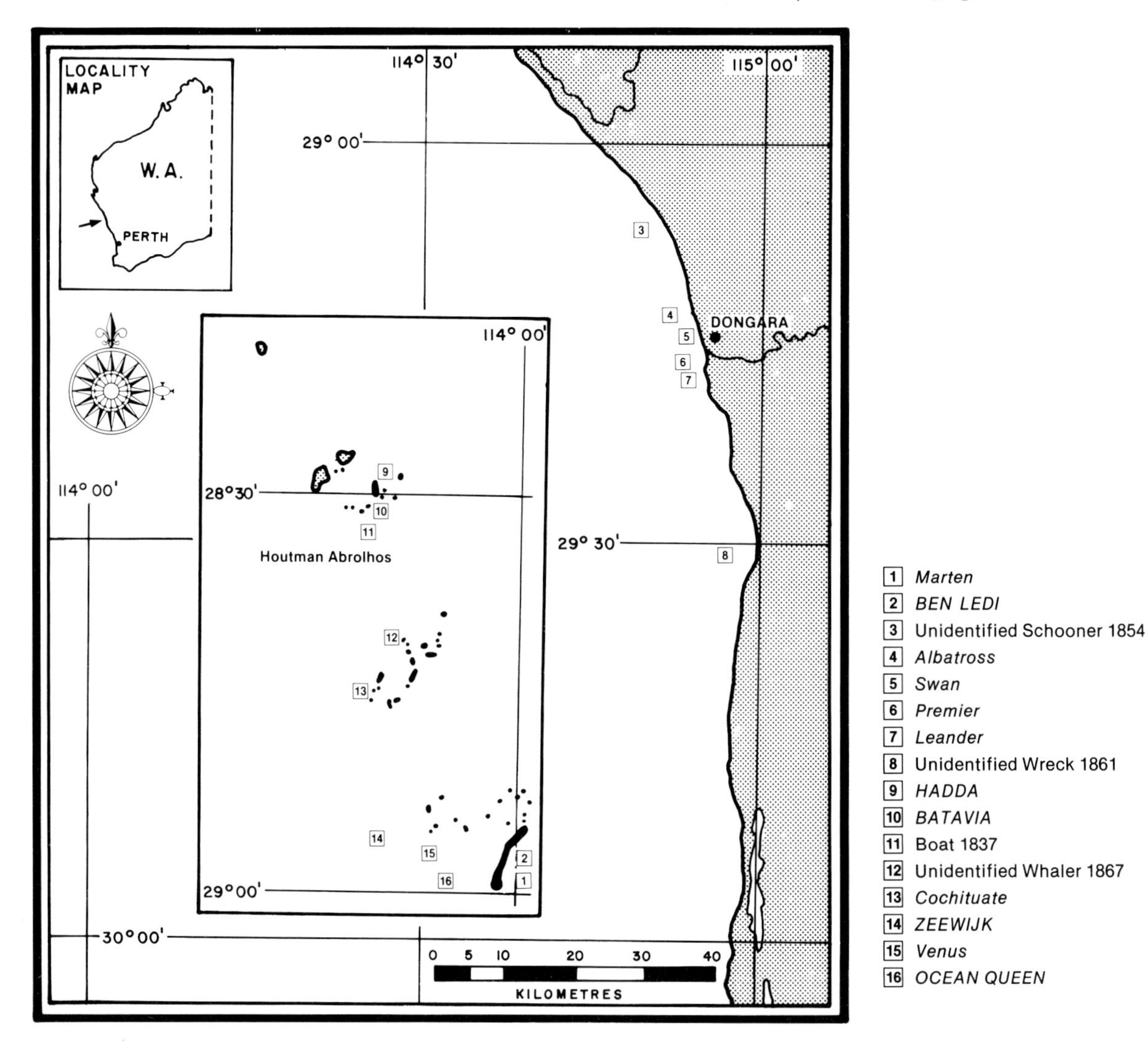
LOCALITY MAP
W.A.
PERTH
114° 30'
115° 00'
29° 00'
DONGARA
114° 00'
28°30'
Houtman Abrolhos
114° 00'
29° 30'
29°00'
30°00'
0 5 10 20 30 40
KILOMETRES
1 Marten
2 BEN LEDI
3 Unidentified Schooner 1854
4 Albatross
5 Swan
6 Premier
7 Leander
8 Unidentified Wreck 1861
9 HADDA
10 BATAVIA
11 Boat 1837
12 Unidentified Whaler 1867
13 Cochituate
14 ZEEWIJK
15 Venus
16 OCEAN QUEEN

ceeded to the Easter Group, 22 kilometres north. One of the salvage party reported:

> Saw the vessel, bearing NW; anchored the cutters under Rat Island, the vessel bearing W by S, distant 4½ miles. At daylight next morning, got the *Speculator* under weigh, and proceeded to the wreck, leaving the *Mystery* at Rat Island. On nearing the scene of the wreck, discovered another reef between the ship and the cutter, almost dry, extending from 70 to 80 fathoms, E to W, the vessel lying about 100 fathoms to the westward, the surf breaking very high the whole way off to the vessel. Launched the dinghy through the surf and got on board the ship; the dinghy immediately capsized, losing two oars; having righted the dinghy, returned to the shore, and brought off the other men. The vessel does not appear to have shifted from her position since the crew abandoned her. The cabin was in great confusion, things lying about in all directions, and on deck just as bad; amidships, in the waist, there was the main-yard, head of the main-mast and main top-mast, but the most wonderful sight was a pig, alive in the stye, which had been there about 70 days, apparently without anything to eat. It was very weak, and could not have long survived. The next thing to be thought of was how to land the provisions, etc, as it was impossible to do so in the boats, as even on the calmest day there is a very heavy sea. Fortunately the mizen-mast was standing, and there being a new hawser on board, a guy was formed by fastening one end of the hawser to the mizen-head, the other being fastened to the reef. The hawser being hauled well taut, a running block with a thimble was placed on it, and worked backwards and forwards; by which means everything was landed from the ship even to the two large anchors and chains. After this the ship took fire, which could only be accounted for in consequence of a cask filled with sand that was used as a temporary galley by the cutter's crew igniting and firing the deck of the vessel. There was nothing of any value remaining when the fire broke out, so it did not much matter.[4]

The *Mystery* trans-shipped her cargo to the *Favourite* and sailed to Shark Bay for guano, while the *Speculator* followed down the coast to collect the *Cochituate*'s two boats and a chronometer left on the beach. Back in Fremantle, the *Favourite* off-loaded '28 casks of beef and pork, peas, bread, flour, hams, the wonderful pig, 13 spars, a set of sails, and a variety of other valuable articles, such as blocks, etc'.[5]

Some readers must have doubted the veracity of the pig story, because the *Perth Gazette* added the following details 2 days later:

> The story of the *Cochituate*'s pig which survived three months fasting on board that unfortunate ship is confirmed so far as the combined assertions of the wrecking party are concerned, who state that the animal was found lying on its side apparently dead, but that some one fancying he saw it's side heave, stirred the poor creature when it gave a faint grunt, and that it was afterwards brought round by doses of castor oil and gruel; a small white sow was shown to us a few days since at Fremantle as being the animal which was found. A fine black cat was also found and is now in the possession of Mr Wellard.[6]

The *Cochituate* (Official Number 2923) is recorded in the *Mercantile Navy List* as a two-decked 347-ton barque registered at Boston.[7] The vessel was built at Medford in 1848 of oak, pitch-pine and hackmatack fastened with copper, and was owned by J. Bangs and son.[8]

NOTES

1. *Inquirer*, 10 July 1861.
2. *Perth Gazette*, 12 July 1861.
3. Report of John Lee, 9 July 1861, Police Records Acc. No. 129, Battye Library.
4. *Inquirer*, 2 October 1861.
5. *Ibid*.
6. *Perth Gazette*, 4 October 1861.
7. *Mercantile Navy List*, 1861.
8. *Registre Veritas*, 1860.

Preston

The schooner *Preston*, built at Preston Point in 1854, first sailed up-river to Perth Water, where her owner, William Owston, announced that he would take the vessel to the north on her maiden voyage, and then to Adelaide for sale.[1]

The vessel was employed between Fremantle and Port Gregory, but at 3 a.m. on 20 September 1855, she ran ashore on Pelsart Island in the Abrolhos. The crew reported that they tried to get her off, but the anchors would not hold, and she drifted further inshore.[2] The cargo of guano was discharged on the 29th, and on 2 October, the vessel was abandoned.

The crew proceeded to Port Gregory in the schooner's boat. There, the *Preston*'s master, John Keefe, obtained the assistance of Captain Sanford and five extra hands, and together they returned to the wreck in the cutter *Nora Creina*. From the 6th to the 11th, they tried unsuccessfully to get the vessel off, and then returned to Port Gregory.

Keefe must have returned again to Pelsart Island over the summer, because in March 1856, the arrivals column of the *Perth Gazette* read: 'on the 9th inst. the cutter *Preston* from the Abrolhos with a cargo of guano'.[3]

The *Preston* continued to operate in the coasting trade until the beginning of July 1861, when she was wrecked at the mouth of the Murray River. Keefe, finding the vessel to be in danger during a gale, deliberately drove her on to the beach in a safe spot. After staying with the vessel for a time, Keefe set out along the beach towards Fremantle for help. But he was later found dead on the road near Rockingham, having apparently died as the result of a fit.[4] On 4 July, Police Constable Stewart reported that the *Preston* was almost a total wreck, but the crew were safe and well.[5]

The *Preston* (Official Number 40471) was built as a 19-ton schooner with the dimensions 13.9 metres by 4.4 metres by 1.7 metres.[6] She had one deck, a square stern, and carvel build.

A wooden wreck, estimated to be around 14.6 metres in length, was found in 1967 by skindiver Jack Yates some 550 metres north-east of the mouth of the Murray River, and reported to the Western Australian Museum.[7] It is possible that the wreck is either that of the *Preston* or the *Alert*, lost on the bar in 1875. Attempts to relocate the wreck seen in 1967 have so far been unsuccessful. (See map on p. 242.)

NOTES

1. *Inquirer*, 8 March 1854.
2. *Inquirer*, 24 October 1855.
3. *Perth Gazette*, 14 March 1856.
4. *Perth Gazette*, 5 July 1861.
5. Report of P.C. William Stewart, 4 July 1861. Police Records Acc. No. 129, Battye Library.
6. Register of British Ships, Fremantle.
7. Western Australian Museum File 206/80.

Unidentified Wreck, Irwin River, 1861

As the shipwrecked seamen from the *Cochituate* trudged southwards, they saw on the beach the yards of a vessel that they estimated would have been between 400 and 500 tons. They also saw candles strewn along the coast. The story was reported in the *Inquirer* in July 1861.[1] In October, the mystery had not been solved, and the *Inquirer* reported:

> There is little doubt that there was another wreck, as the spars and top-sides of a large vessel have been seen, on the beach 12 miles to the South of the Irwin.[2]

It seems that the Colonial Government was quite concerned about the wreckage. In December, Governor Arthur Kennedy wrote to the British Government with reports of a piece of wood and an anchor buoy, and copper bolts were sent to the Admiralty for analysis.[3]

The sailors may have seen the remains of a recently wrecked ship. However, the stated position matches reasonably that of the *Leander* wreck of 1853, and it could be that the stormy weather had scoured up some of the remains of this vessel or that of the unidentified wreck reported in 1851. (See map on p. 44.)

NOTES

1. *Inquirer*, 3 July 1861.
2. *Inquirer*, 2 October 1861.
3. A. G. Kennedy to Lord Newcastle, 25 December 1861, Governor's Despatches No. 128, Battye Library.

Pilot's Boat, 1862

On the night of 25 January 1862, two probation men escaped from the convict depot at Port Gregory, taking an old pilot boat, equipped with oars, that had been sent up from Fremantle for the service of the Government. But the proper management of the boat required at least six hands, and the escapade ended in disaster.

Police who were sent along the coast in search of the men found one body on the beach, quite naked and badly mutilated by sharks. It appeared that the boat had gone up on a reef and the men had stripped in an attempt to swim ashore.[1](See map on p. 53.)

NOTE

1. *Inquirer*, 12 February 1862.

Fitzgerald

The schooner *Fitzgerald* was built on the Swan River at Perth in 1861. In July of that year, the vessel was taken to Champion Bay under the command of Edward Brown, and she made other coastal voyages.[1] But the *Fitzgerald* did not survive for long. She was being brought over to Fremantle from Garden Island on 15 July 1862 when she struck some rocks and went down. The crew were saved in the *Fitzgerald*'s boat.[2] (See map on p. 242.)

Brown related the events at the court of inquiry:

> She struck on a rock north-east of the Fish Rock about 200 yards about 7 p.m. It was quite dark at the time. At sunset I was to windward of the rock upon which I was wrecked, and stood to the westward to give the Fish Rock a wide berth, knowing at the time that a strong current was running to the southward. I stood on this tack about half an hour and then went about, and stood north by east, then tacked again to the westward for about three quarters of an hour, and then tacked and stood towards Fremantle, the light being sometimes on the lee bow and some-

> times ahead, all hands were on deck and looking out as well as myself. I went twice to the foreyard to see if I could see any breakers; the wind was at this time north west. When she struck the rock I was forward and looking out, and there was no appearance of danger that I could see. I at once threw the sail back to get off the rock; she filled at once and went down; we had just time to get into the boat and save ourselves, three sailors and myself, and three passengers.[3]

Wreckage and part of the cargo were later washed up on the shore near Woodman's Point, where a party of convicts were employed collecting timber that had been washed away from the South Jetty. When the men failed to return to the gaol in the evening, the Police started along the beach and found the convicts drunk.[4] It was presumed that they had found bottles of spirits washed up from the *Fitzgerald*.

The 24-ton *Fitzgerald* (Official Number 36543) was built with two masts and a square stern. Her dimensions were 15.7 metres by 4.5 metres by 1.6 metres, and she was owned by John Mews.[5]

NOTES

1. *Inquirer*, 17 July 1861.
2. *Inquirer*, 23 July 1862. The *Inquirer* called her the *Lady Fitzgerald* in this article, but the Register of British Ships lists her as the *Fitzgerald*.
3. *Inquirer*, 1 August 1862.
4. *Inquirer*, 23 July 1862.
5. Register of British Ships, Fremantle.

Lapwing

In July 1862, the *Inquirer* reported:

> that several pieces of wreck have been washed up on the beach at Bunbury, in the shape of a figurehead of a ship, a stern board bearing the name *Lapwing* and a portion of a windlass.[1]

Several American whalers bore the name *Lapwing*. One of these, a ship of 432 tons owned by E. Jones, of New Bedford, visited Western Australian ports in 1854, 1858 and 1863. It was reported as arriving at the Vasse seeking more crew on 6 February 1863 under the command of Captain George Soule. She had been 32 months out from her home port and had 1,600 barrels of sperm oil and eighty-five barrels of other whale oil on board.[2]

Another vessel, the 502-ton *Lapwing* (Official Number 11364) of Baltimore, is not recorded as having visited Western Australian ports.

W. J. Dakin, in his *Whalemen Adventurers*, recalls an interesting legal case resulting from the New Bedford *Lapwing*'s 1858 visit and reported in a New Bedford newspaper in 1859. Dakin writes:

> It bears the date of 1859 and refers to a case at law where a 'native of the colony and subject of the queen' brought an action against a Captain Cumisky of the whale ship *Lapwing* of New Bedford on the grounds that aliens were not allowed to whale in the bays of the colony. The plaintiff set out that he had made it his business to catch whales in Fremantle Bay with boats from the shore, and that in November 1858 the defendants had frightened and driven away a whale he was pursuing. The jury gave the W.A. colonist a verdict for £300 damages. A new trial was demanded, and it came out that although the plaintiff had chased the whale for 3 miles he was never quite near enough to throw his harpoon. He said the defendants then sent a boat in pursuit, and by following in its wake they did the most effective thing to frighten it, *'because a Right whale can only see behind it!'* No other witness agreed about the whale only being able to see behind, since the eyes are on the sides of the head. The judge held that the verdict was defective, and quoted from certain authors at length, finally deciding that the whalers had the right to enter the bays and that the Americans were alien friends to whom, by the laws of England, almost all the privileges of natural born subjects were granted.[3]

No other information has come to light to indicate whether the *Lapwing* sternboard was lost as a result of the ship being wrecked or simply as a result of lack of maintenance.

NOTES

1. *Inquirer*, 23 July 1862.
2. *Perth Gazette*, 13 February 1863. See also A. Starbuck, *History of the American Whale Fishery from its earliest inception to the year 1876* (Waltham, Mass., 1878).
3. W. J. Dakin, *Whalemen Adventurers* (Sydney, 1963), p. 118.

African

The ship *African* left Champion Bay for Fremantle on 1 January 1863. She carried on board 522 tons of copper and lead ore, and upwards of 300 bales of wool. Like the *Leander*, the *African* would have visited the Geraldton area for heavy ore with which to trim the wool cargo. It is likely that the *African*'s master chose to avoid Port Gregory because of the deep draft of his vessel, or because of the shipping casualties that had occurred there. The ore would have been lightered from Port Gregory to Champion Bay for the *African*.

Passengers later stated that the ship's pumps were set to work soon after she got outside the harbour, and a heavy gale from the south-west kept the crew busy trying to clear the bilges. After contending with heavy winds until the 4th, the course was changed for a return to the Bay. At 10 p.m., the *African* touched a reef, which Captain Gibson estimated was 19–22 kilometres south of Point Moore. The next morning, she came into harbour and ran aground 140 metres from the jetty head in 5 metres of water.[1] She had at that time 2 metres of water in her hold, but this was soon reduced to 1.3 metres by the pumping efforts of a party of prisoners from the local gaol.

All the cargo was unloaded, but the leak continued. An inquiry into the accident concluded that the damage was caused by the vessel striking the reef,

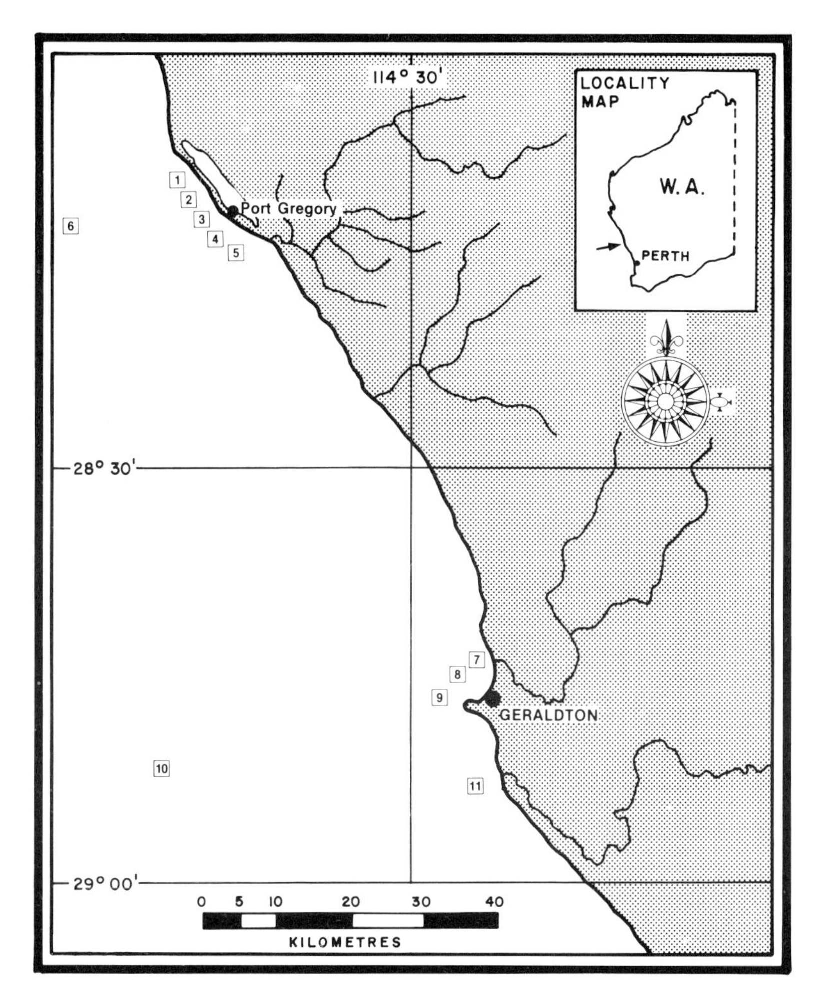

1 *XANTHO*
2 *Favourite*
3 *Sara*
4 *Mary Queen of Scots*
5 Pilot Boat 1862
6 *Star*
7 *AFRICAN*
8 Unnamed Boat
9 Unnamed Lighter
10 *Flying Foam*
11 Unidentified Ship 1851

and not from any previous defects in the hull, or inattention to the loading of the cargo.[2]

Captain Gibson had the hull surveyed and decided that the *African* was not in a fit state to proceed on her voyage, that she could not be repaired at Champion Bay, and that he had insufficient money to send her elsewhere for repairs. So he abandoned the vessel as a constructive total loss and gave instructions to Lionel Samson, the auctioneer, to sell her. He disassembled the rigging, and even such fittings as brass scuttles and the capstan, on the basis that they would bring more if sold separately.[3]

The merchant Luke Leake purchased the hull, rigging and fittings for about £1,000, intending to get the ship repaired in some Asian port.[4] But he did not pay cash at the time of purchase and later repudiated the bargain successfully in court.

Meanwhile, on 23 March, the *African* dragged her anchor for 5 kilometres to the north, stopping off the mouth of the Chapman River just in time to avoid a reef.[5] Six weeks later, the *African* still held together, lying between two reefs with her stern to shore and two masts standing.[6] Samson, who was also a prominent merchant, was interested himself in purchasing the wreck and wrote to Geraldton merchants Scott and Gale, suggesting a joint investment. They declined:

> We have by us yours of 26 ult [May] in which you ask us to join you in purchasing the hull of the *African* or to give you our ideas on the subject. As for us joining in the purchase we must decline your offer, as we do not think money is to be made out of her at £415—where she now lies and at this time of year. You need not let any one know our opinion unless you like, and the sooner she is sold the better: she lies about 200 yds from the shore with her bow to the Beach, and listed to Port, the mainmast and foremast are standing—The things saved from her are about 6 water c [closet] fittings and a lot of cabin fittings carefully taken to pieces. The rope, sails, spars etc on board of her belong to us, as they were put on board to try and save the vessel and bring her to her former anchorage.[7]

A second auction was held on 22 June, and Samson bought the wreck for £70.[8] A shipwright named William Garrard bought some of the timbers and built from them the cutters *Albatross*, *Lass of Geraldton* and *Mary Ann*.[9]

The wreckers did not take all of the ship. In 1977, divers Alan Paterson and Gary Orgill reported finding material 300 metres north of the Chapman River at latitude 28° 43.4′ south, longitude 114° 37′ east. Scattered timbers, iron fittings, yellow metal bolts and sheathing, and lead ore lie over an area some 50 metres by 20 metres between two reefs. The finders were given a reward of $100 by the Western Australian Museum and the site was gazetted under the Historic Shipwrecks Act as the Sunset Beach unidentified wreck.

The *African* (Official Number 23091) was a 780-ton ship-rigged vessel owned by Gibson and built at Sunderland in 1853.[10] Her dimensions were 48 metres by 10 metres by 6.5 metres. She was copper fastened and, in 1860, was sheathed with yellow metal.

NOTES

1. Minute by Surveyor General, January 1863, in Letters Forwarded to Officials, Lands and Survey Dept, pp. 433–4, Acc. No. 229, No. 43, Battye Library.
2. *Inquirer*, 4 February 1863.
3. Samson versus Leake, *Inquirer*, 13 May 1863.
4. *Inquirer*, 4 March 1863.
5. *Perth Gazette*, 3 April 1863.
6. *Perth Gazette*, 13 May 1863.
7. Scott and Gale to Samson, 3 June 1863, Samson Collection Acc. No. 2169A/2, Battye Library.
8. Account of Sales of the Wreck of the *African*, 22 June 1863, Samson Collection Acc. No. 2169A/26a, Battye Library.
9. *West Australian*, 4 July 1936.
10. *Lloyds Shipping Register*, 1863.

On board the *Zephyr*.

Zephyr

In April 1864, it was announced that the British shipowner Robert Habgood had purchased a former China tea-trade clipper for the Western Australian wool trade.[1] The 409-ton barque *Zephyr* was to land British cargo at Fremantle, convey machinery to the Geraldine Mine at Port Gregory, load lead or copper ore as a paying ballast, and then cram her hold full with wool for the London market.

The schedule ensured full paying cargoes both ways between Britain and Australia. The only problem was that Port Gregory, source of the paying ballast, was not a secure port.

The *Zephyr* entered Port Gregory on 28 September 1864 with a fair wind. She had just anchored when the breeze freshened and she bumped on a sandbank where, instead of 5.5 metres of water as stated on the chart, there was only 2.4 metres. The bank had shifted.[2] The crew tried to warp the *Zephyr* off, but she parted from both anchors. The vessel went twice her length up the sand, the ground being quite level. It was thought that she would become a total wreck.[3]

Everything was taken out of the ship to lighten her. During this operation, a boat carrying eight men capsized, and three men were drowned. A high tide on 12 October provided the opportunity to refloat the *Zephyr*, and she returned to Fremantle with little damage. The *Zephyr* was to remain in the wool trade for many years, braving other coastal outports further north to collect wool. In later years, at least one insurance company refused to cover ships calling at Port Gregory at any time of year, or Port Irwin in the winter.[4]

NOTES

1. *Inquirer*, 20 April 1864.
2. *Inquirer*, 12 October 1864.
3. Report of P.C. Thomas Alleby, Geraldton, 29 September 1864. Police Dept. Records, Acc. No. 129, Battye Library.
4. Southern Insurance Company, Policy No. 258. Habgood Collection, Acc. No. 813A, Battye Library.

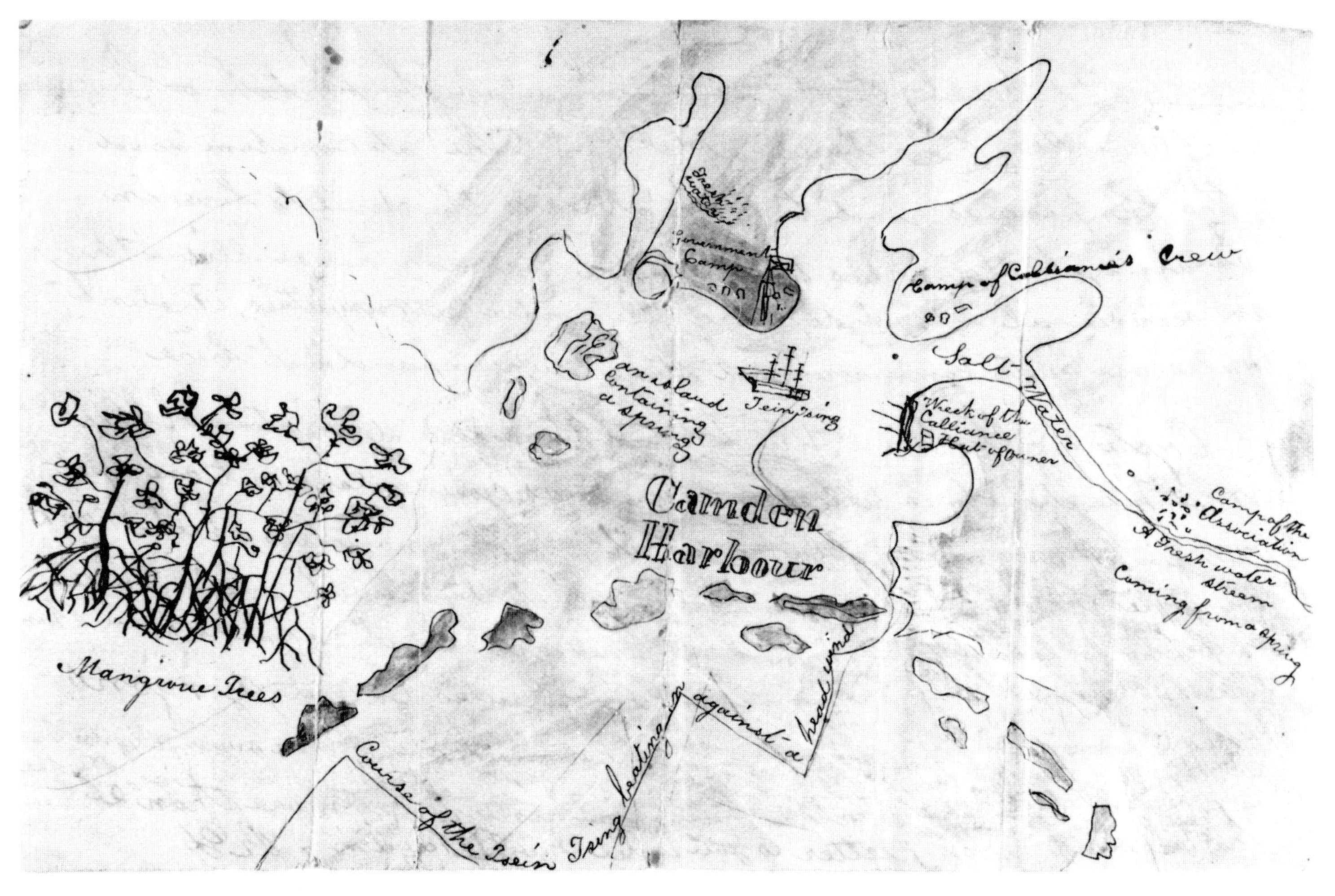

Contemporary sketch map of Camden Harbour, showing the *Calliance* wreck.

Calliance and Dutch Gun-Boat

In 1864–1866, a Melbourne syndicate promoted a colonising expedition to open up the tropical country in the vicinity of Camden Harbour in the West Kimberley region. The project was organized without any real knowledge of the country, and when they had had an opportunity to judge the land for themselves, most of the dispirited settlers eventually moved south to the newly opened pastoral areas. The events surrounding the loss of the *Calliance* show something of the hardships endured by the settlers in this brutal environment.

The 822-ton ship *Calliance*, bringing intending settlers and their livestock from Melbourne, was approaching the vicinity of Camden Harbour on 22 December 1864 when she struck an unmarked reef near Adele Island. Captain Brown had his crew throw overboard 130 tonnes of ballast, 120 bales of hay and 6,800 litres of water before the ship came off to limp into the harbour. Their arrival on Christmas Day was surrounded by gloom:

> Sunday 25th. At 0.30 a.m. came to an anchor off the entrance of Camden Harbour. At 6 a.m. weighed and proceeded into Camden Harbour. At 10 a.m. anchored in 6 fathoms, 25 days from Port Phillip, with 120 deaths out of the 2,700 sheep taken on board. Air close, and sun very hot; ther. 89 deg. About 5 p.m. Mr Hart, passenger, found insensible, having had a sunstroke; at 7 p.m. died; buried him on Sheep Island, the Rev. Mr Tanner reading the burial service at 7 a.m. 26th. Found, from the report of the passengers previously arrived by the *Stag* and *Helvetia*, there was apparently very little food, and no water within a few miles of the ship for the sheep . . .[1]

By 31 December, the crew had unloaded alive 2,500 sheep, fourteen dogs, eight goats and sixty-two passengers. On 4 January 1865, the *Calliance* was moved to the entrance of a bay and kedged towards shore, with a view to careening her. The tide rose 7.6 metres that day.[2] But the next day, the sea breeze blew the *Calliance* ashore, and as the tide fell, the ship heeled over to starboard amid frequent loud cracks as timbers broke. She was perched on two large rocks and the keel had been bent into two arches. The tide flowed as fast inside the ship as it did outside. (See map on p. 152.)

On the 17th, the *Calliance*'s cutter set out for Coepang in Timor to try to procure a vessel to take the *Calliance* survivors away from Camden Sound. The cutter, under the command of the second officer, Mr Clark, with Captain Edwards (one of the settlers) and three crewmen, reached Timor in 5 days. At the request of the Governor of Timor, a small schooner, described as a Dutch gun-boat, was loaned to the party by a Mr Drysdale of Coepang.[3]

The two boats then got under weigh on the return voyage. About 130 kilometres out from Camden Sound on the morning of 8 February, a heavy squall struck and upset both the schooner and the cutter, which had been sailing a little behind. The men clambered onto the overturned hull of the cutter, and one of the crew dived under her to cut away the mast. Then they were able to right the cutter, and, using a bundle of hats that floated out of the schooner, they bailed the cutter dry.

One of the men dived under the overturned schooner and kicked out the cabin window, hoping to find the body of the missing Captain Edwards. He managed to catch hold of the dead captain's hair and drag him out through the window. They then tried to retrieve from the cabin some provisions and money belonging to the captain, but the gathering sharks obliged them to set sail, having saved nothing but the hats and a few floating coconuts.

Three days later, they reached New Island (off Camden Harbour) and got water. They intended burying the captain there, but were frightened off by a party of Aborigines and were compelled to throw the decomposing body overboard before they got into harbour.

The settlers were preparing the *Calliance*'s launch for another voyage to Coepang when the barque *Tien Tsin* arrived with a party of Government officials on 16 February. The launch set off for Timor on 8 March, with a fortnight's provisions. The *Tien Tsin* left soon afterwards. Captain Brown, of the *Calliance*, wrote:

> 19th. All on board the *Tien Tien*, including 53 passengers of the Camden Harbour Pastoral Association, now extinct, and who represent three fourths of the capital, and four fifths in number of persons, who formed this most unfortunate and misled body of men.[4]

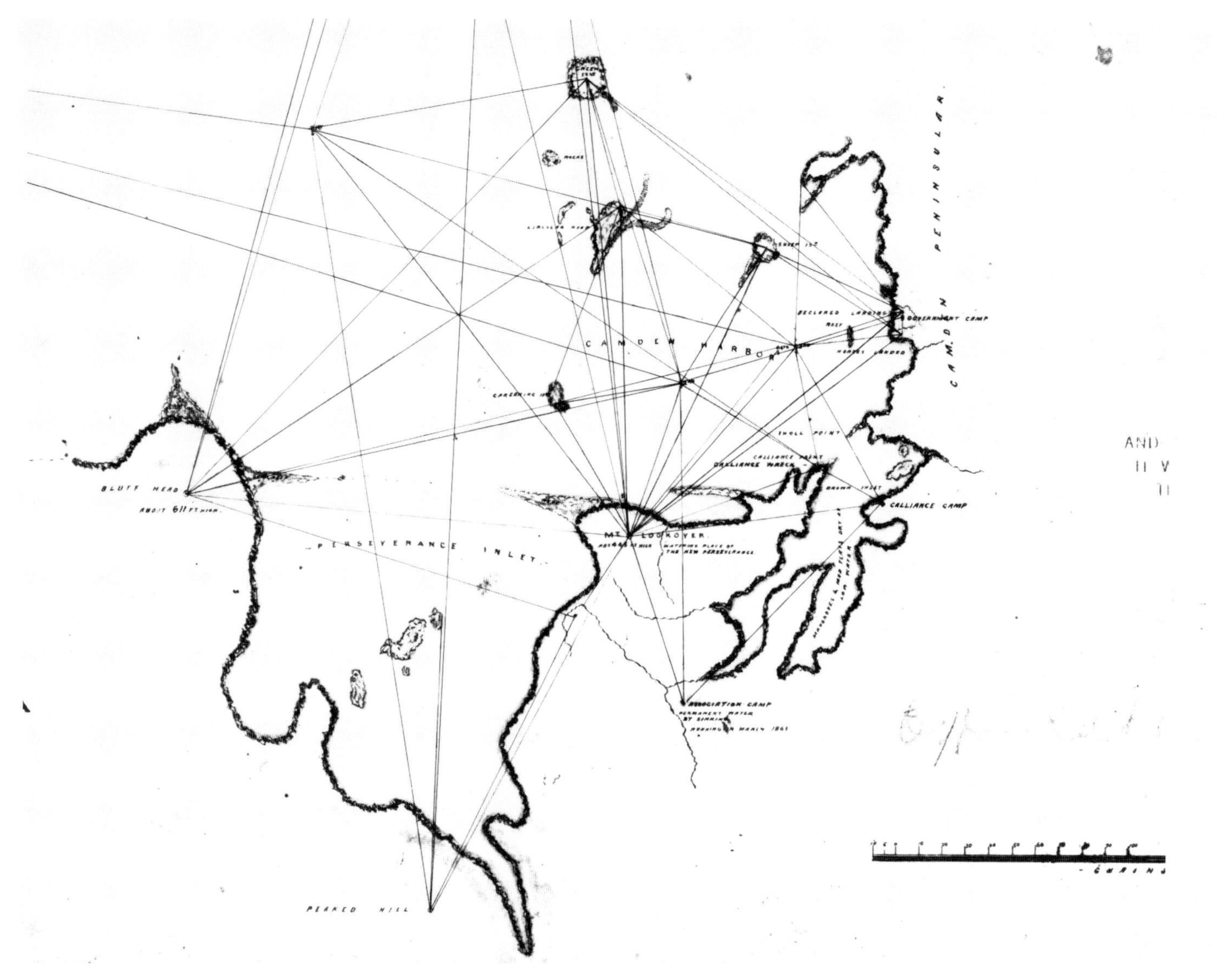

A later chart of the same area, placing the wreck more accurately.

Among those who chose to stay were the purchasers of the wreck of the *Calliance*, Messrs Anderson and Co. Jefferson Stow, who visited Camden Sound in June 1865, saw the three men at work:

> On leaving the landing at the camp we sailed over to the wreck of the *Calliance*. Here several of the Victorian settlers who had purchased the wreck had their tents erected on the shore, and were busy in preparing to burn the hull as the only way to get the copper from her. We purchased a cable, and soon after sailing, an enormous column of smoke arose from the wreck. The *Calliance* was on fire as if in honour of our departure.[5]

An expedition from the Western Australian Museum visited Camden Harbour in 1978 and found the sparse remains of the *Calliance* wreck, consisting of hearth bricks, iron nails and yellow-metal sheathing.[6] But what the wreckers left, the heavy tides have dissipated. The wreck site, the wreckers' campsite, the survivors' campsite, the Association main campsite and the graves on Sheep Island all have the protection of the Maritime Archaeology Act. Ian Crawford, in his *Art of the Wandjina*, described wreckage on New Island, consisting of brass pipes, pieces of iron, broken china ware and part of a compass ring.[7] The 1978 Museum party found iron frames and 1-inch-diameter bolts. The bolts suggest a large vessel such as the *Calliance* rather than a small vessel such as the lost Dutch schooner.

The *Calliance* (Official Number 26480) was an 822-ton ship-rigged vessel built at Sunderland in 1854, and owned by John Cereswell of Sunderland.[8] Her dimensions were 48 metres by 9.7 metres by 6.4 metres. She was sheathed with yellow-metal to the wales in 1854.[9]

NOTES

1. Journal of Captain Brown, in *Inquirer*, 26 April 1865.
2. *Ibid*.
3. Journal of Trevarton Sholl, 17 February 1865, QB/SHO, Battye Library.
4. Journal of Captain Brown.
5. Jefferson Stow, *The Voyage of the Forlorn Hope*, 1865 (Adelaide 1981).

6. Scott Sledge, *Report of Wreck Inspection North Coast, 1978.* Western Australian Museum.
7. I. M. Crawford, *The Art of the Wandjina* (Melbourne, 1968), p. 79.
8. Board of Trade Wreck Register, 1864, National Maritime Museum, England.
9. Lloyds Survey Register, No. 5387, National Maritime Museum, England.

Archaeologist Scott Sledge examines bricks on the site of the *Calliance* wreck.

New Perseverance and Government Pinnace

The 105-ton schooner *New Perseverance* was built on the Swan River by Captain Owston in 1857. Samuel Brakes, the schooner's first owner, ran into a spot of bother with the vessel before she had even commenced her maiden voyage. He was just getting the *New Perseverance* under weigh for South Australia when the Water Police came on board and found slipway blocks that were alleged to have been stolen from the vessel *Les Trois Amis*. Brakes was also charged with having concealed on board a runaway ticket-of-leave holder, and was imprisoned.[1] So when the *New Perseverance* sailed, under the command of Mr Adams and with forty-five passengers, the owner was not on board. The master Brakes chose for the *New Perseverance* did not look after her adequately. She was badly damaged during the voyage to Adelaide, the repairs costing upwards of £800. The unsupported Mrs Brakes sought the assistance of the Governor, claiming that she had been compelled to auction the vessel, but had realized less than £800 on the sale.[2] The vessel was bought at this bargain price by George Shenton.

The *New Perseverance* was sent by Shenton on several voyages to the north of the Colony. In 1864, she took a party of surveyors under the leadership of Frederick Panter to explore the country around Camden Harbour and Roebuck Bay (later the site of Broome). On the night of 29 March, she struck the then unmarked Treacherous Reef near Capparelli Island in the Buccaneer Archipelago, and was nearly lost.[3] Plans to settle the plains near Roebuck Bay were jaded when Panter and his companions Harding and Goldwyer were killed by the Aborigines the following year.

In November 1866, the *New Perseverance*, now owned by William Owston, was employed to transfer stock from Roebuck Bay to Nickol Bay, close to the later site of Roebourne. The vessel successfully carried out the loading at Roebuck Bay and returned to Cossack, near Nickol Bay. However, on entering Butcher's Inlet on 20 December, the wind fell and the vessel was obliged to come to anchor near a patch of rocks. She then drifted onto a rock and knocked a large hole in her side. By continual pumping, she was kept afloat until beached. The Government's pinnace foundered and was wrecked while assisting the *New Perseverance*. (See map on p. 118.)

Among the losses when the *New Perseverance* was wrecked were a large section of the Roebuck Bay police quarters (presumably disassembled and lying in the hold of the vessel), and some timber belonging to the Roebuck Bay Company. Most of the crew obtained employment in the district, the hapless remainder, with the ex-master of the *New Perseverance*, the staff of the Roebuck Bay Company, and the police and pensioners brought from Roebuck Bay, taking passage on the ill-fated *Emma*.[4]

The *New Perseverance* remained on the beach, and her registry was closed in 1868. In 1869, she was used as a storage hulk for pearl shell. In December 1870, a cyclone shifted the wreck further up the creek and swept away all her deck houses.[5] In later years, the story was told that a tidal wave drove her so far up on the land that she was permanently dry, and that the enterprising George Seubert used the snug confines of the hull as a taproom when he secured a licence for the sale of liquor in Cossack. Seubert, a pearler and publican, was later obliged to become a boarding-house keeper at Fremantle after his leg was amputated.

The *New Perseverance* (Official Number 40476) had one deck, two masts, a figurehead of a woman and a square stern. Her dimensions were 25.9 metres by 6.3 metres by 3 metres.[6]

NOTES

1. *Inquirer*, 30 January 1857.
2. Lionel Samson to Col. Sec., 22 July 1857, C.S.R. 371, fol. 244. The Register of British Ships, Fremantle, indicates that Brakes had already sold the vessel to Lionel Samson on 26 January.
3. Report of the Surveyor General, 23 July 1864, Letters Forwarded to Officials, Lands and Surveys Department, Acc. 229, Battye Library.
4. *Perth Gazette*, 12 July 1867.
5. *Inquirer*, 15 February 1871.
6. Register of British Ships, Fremantle.

Brothers

The 16-ton cutter *Brothers* was built on the Vasse River in 1854, and was owned by George Chapman of Fremantle, who employed her in the coastal trade.[1]

The Registry states that, in 1863, the vessel was grounded on a sand-bar on the Murray River, and finally abandoned after all attempts to get her off had failed.[2] However, she must have been refloated, because her new owner, contractor William Spencer of Bunbury, advertised her for freight or passage in 1866.[3]

The *Brothers* sailed for Roebourne on 19 February 1867 under the command of J. Williard, carrying one passenger and four crew. They were never seen again.

The pearler and pastoralist Charles Tuckey offered an explanation that may or may not help to explain where the *Brothers* was wrecked.[4] He said that on a voyage he made in 1876 from Port Walcott to Fremantle, an Aborigine, whose tribe inhabited the country to the south of North West Cape, told him about three wrecks in the area. Tuckey took it that the vessels were the *Occator*, wrecked in 1855, the *Emma*, wrecked in 1867, and the *Brothers*. He took his vessel inshore as far as he considered prudent after rounding the Cape, and made out distinctly, with the aid of his telescope, the ribs of a vessel lying on the beach. The place was situated about 160 kilometres south of the Cape, the area today occupied by Warroora Pastoral Station. Tuckey supposed that the *Brothers* and the *Emma* might have been wrecked while standing in at night. (See map on p. 30.)

Lockier Burges in his reminiscences stated that the *Brothers* was lost off Dongara, but there appears to be no evidence to corroborate this.[5]

The *Brothers* (Official Number 40481), of dimensions 11.6 metres by 3.9 metres by 1.7 metres, had one deck and a break, a square stern and carvel build.[6]

NOTES

1. Register of British Ships, Fremantle.
2. Ship Registry Transcripts 1832–1862, Acc. No. 387.2/1765P, Battye Library.
3. *Inquirer*, 1 August 1866.
4. *Inquirer*, 19 January 1876.
5. L. C. Burges, *The Pioneers of the Nor'-West, Australia* (Constantine and Gardner, Geraldton, 1913), p. 32.
6. Register of British Ships, Fremantle.

Emma

The schooner *Emma* was built in Lowestoft, Suffolk, in 1859, and was bought by the Western Australian pastoralist and merchant Walter Padbury in 1865. Misfortune pursued the *Emma* from the date of her arrival on the Western Australian coast. On her first voyage to the North-West she lost a man overboard and lost an anchor off the De Grey River. At Champion Bay, she collided with the jetty, causing damage. At Cossack, she went aground. On her next voyage north from Fremantle, she went aground on the Abrolhos and her cargo of sheep had to be off-loaded on the Islands before being taken into Champion Bay. Passenger Lockier Burges later reminisced:

> We left Fremantle in a bustle, with trusses of hay piled on the main deck so in fact everything was in the wrong place. If rough weather had come on suddenly we should have been in a fix. To make matters worse a large compass that had belonged to the *Calliance* (wrecked at Camden Harbour) was placed in the *Emma*. The *Calliance* was three times the tonnage of the *Emma* , so consequently the compass would not work correctly in such a small vessel. The result was that the next night after leaving Fremantle, we found ourselves high and dry on the south end of the Abrolhos . . . . We had only one boat—such were the conditions in these olden times, that we were allowed to go to sea with only one boat, and that not even a lifeboat, but an old tub, built of jarrah, which would go down like a stone if she capsised.[1]

A schooner, similar to the *Emma*, stranded on the beach at Fremantle in 1862.

Returning to Fremantle, the *Emma* struck a sand-bar south of the old jetty and was dismasted. She was refloated, but the next north-wester sent the schooner back onto the beach.[2]

Padbury paid another £100 to have her refloated, after which she was fitted up with new rigging and thoroughly renovated for another voyage to Cossack.

At Cossack, Captain Badcock and his seven crew loaded wool belonging to the Roebuck Bay Company, and several tonnes of pearl shell. Thirty-four passengers embarked for Fremantle, and the *Emma* left Roebourne on 3 March 1867. It was Badcock's intention to obtain more provisions from Fremantle. The average duration of colonial coasting voyages from Fremantle to Port Walcott was 30 days, with 50 days for the return voyage.[3] The *Emma* must have been a swifter vessel than the average coaster. A previous voyage up the coast had taken 19 days, so it was expected that she would be back in Roebourne by the end of April.

The *Emma* did not return. Nor did the *Brothers*, which was also expected with supplies. Food ran short in the settlement, and in May, as a desperate measure, Robert Sholl, the Government Resident at Nickol Bay, sent a party overland to Champion Bay (about 1,000 kilometres as the crow flies) to seek relief.[4] On their arrival, a supply vessel was sent north.

Speculation as to the fate of the *Emma* and those on board was varied. Captain Tuckey, who had sailed from Fremantle in March and arrived safely at Roebourne, thought he discerned the mast of a vessel on the shore while passing Dirk Hartog Island.[5] Sholl recalled that the *Emma* had been very lightly ballasted, having 25 tonnes of iron ballast and only a few tonnes of cargo and passengers' luggage.[6] He also noted that her mainmast was defective, and speculated that the vessel might have been either upset in a squall or dismasted to float helplessly. Captain Badcock had told Sholl that he would sail round Ritchie's Reef (now named Tryal Rocks, north of the Monte Bello Islands) and then keep close along the shore if possible. Sholl also wrote that 'she had a good slant of wind which ought to have carried her round the North West Cape'.[7]

The speculation was revived in 1876 by Charles Tuckey (see *Brothers* entry) who claimed that an Aborigine from a North West Cape tribe had told him the following story:

> A long time ago (about ten years he described) a ship was wrecked near North West Cape; the passengers landed, at night, in the boats, and as they had no means of defending themselves the natives had no difficulty in making them prisoners. There was a large number of persons, and amongst them were some females [no women are listed among the *Emma*'s passengers]. The natives were not 'sulky' with them, but nevertheless they killed and ate all of them, the narrator partaking of some of the flesh.[8]

It is possible that Captain Tuckey's account of the Aborigines' behaviour was influenced by his attitude towards Aborigines generally. In later years, as master of the *Argo*, he engaged in rounding up and kidnapping Aborigines for forced labour in the pearling industry.[9]

In the 1930s, a stockman on Warroora Station, about 160 kilometres south of North West Cape, is said to have found a small iron cannon on the beach at the front of Maggie Cliffs. On a hill in the adjacent dunes lay a stone cairn.[10] Neither can be located today. (See map on p. 30.)

Diver Serge Katarshi reported in 1979 the finding of a wreck in the Coral Bay area, and inspection by the Museum recently suggests that the site is that of the *Emma*.

The 116-ton schooner *Emma* (Official Number 25291) had two masts, one deck, a round stern and a shield head.[11] Her dimensions were 26.1 metres by 6.2 metres by 3.4 metres.

NOTES

1. L. C. Burges, *The Pioneers of the Nor'-West, Australia* (Constantine and Gardner, Geraldton, 1913), p. 29.
2. *West Australian*, 22 March 1886.

3. *Government Gazette*, 20 February 1872.
4. *Perth Gazette*, 12 July 1867.
5. *Ibid*.
6. *Inquirer*, 28 August 1867.
7. *Perth Gazette*, 12 July 1867.
8. *Inquirer*, 19 January 1876.
9. Su-Jane Hunt, The Gribble Affair, History honours degree thesis, Murdoch University, 1978, p. 38.
10. B. Parker, 'Northern Cannon Quest', *Underwater Explorers Club News*, 11: 1 (1971), p. 1.
11. Register of British Ships, Fremantle.

Lass of Geraldton

The schooner *Lass of Geraldton* was built at Champion Bay in 1865. The *Inquirer's* correspondent gave her the usual launching day compliments, describing her as 'a very smart looking craft, a regular clipper in appearance'.[1] But he conceded that 'of course her lines do not please everyone, some considering she has overhangs too much in the stern and that she is too sharp forward'.[2] It seems that her looks did not improve with age. Lockier Burges later reminisced:

> The *Lass of Geraldton* was a strange looking craft. Being first built as a cargo boat, she was subsequently added to aft in order to carry passengers. If not weighed down heavily forward she had a habit of cocking her bows straight up, then making a dive forward and burying herself up to the bulk heads in every sea. The captain and myself had to move about 140 bags of wheat and two tons of lead ore forward to steady her.[3]

The vessel joined the coasting trade on the west coast. On 25 March 1867, she set out on such a voyage from Fremantle to Bunbury, under the command of Henry O'Grady, with five crew and two passengers, including the pioneer merchant and mine-owner George Shenton, a part owner of the vessel. The *Lass of Geraldton* cleared out in ballast, consisting of about 15 tonnes of sand, and 2 tonnes of flour subsequently put on board.[4]

At about 1 p.m. the same day the, vessel became unmanageable during a squall, heeling right over and gradually going down. Four crewmen and the two passengers drowned, while O'Grady and an Indian crewman swam ashore. O'Grady later wrote to his father:

> I held poor Mr Shenton on a hatch for four hours, and saw that he was dead before I left him, and I also saw two of the crew dead.[5]

Search parties found on the beach the hatches and about ten bags of flour from the *Lass of Geraldton*. The wreck itself could be seen from the shore, about 20 kilometres south of the mouth of the Murray. It was between 3 and 5 kilometres out to sea in about 14 metres of water, with one mast protruding. A party who examined the wreck from a whale boat expressed the opinion that she could be refloated, and it was reported later that two coasting craft had been despatched to attempt to raise her.[6] However, her registry was closed in December of that year. (See map on p. 242.)

The *Lass of Geraldton* (Official Number 52231) was a 32-ton schooner with one deck, two masts, a square stern, and dimensions of 18.4 metres by 4.8 metres by 2.1 metres.[7]

NOTES

1. *Perth Gazette*, 4 August 1865.
2. *Ibid*.
3. L. C. Burges, *The Pioneers of the Nor'-West, Australia*, p. 30.
4. J. T. Reilly, *Reminiscences of Fifty Years of Residence in Western Australia* (Perth, 1903).
5. *Perth Gazette*, 29 March 1867.
6. Report of Sub Inspector W. Snook, 30 March 1867, Police Files Acc. No. 129, 10/582, Battye Library. See also *Inquirer*, 3 April 1867.
7. Register of British Ships, Fremantle.

Lady Lyttleton

The 178-ton barque *Lady Lyttleton* was on a voyage from Adelaide to Fremantle in June 1867 when she was dismasted by heavy weather off the south coast. The vessel was leaking, so her commander, John McArthur, put in for Albany and threw overboard a quantity of his cargo of flour.[1]

The Lady Lyttleton arrived at King George Sound on 16 June and continued to unload cargo. She was then taken to the south side of the channel at Emu Point and hove down for careening. But the vessel slipped on the bank while being hauled down, and sank.[2] In July, the *Inquirer* could report:

> A vessel from Port Adelaide . . . has gone down at King George's Sound, after having encountered and been driven about in the most tempestuous weather.[3]

Captain McArthur hired a lighter for the stowage of salvaged cargo, some of which was sold in August. The *Emily Smith*, from Adelaide, commenced loading much of the remainder to be taken to Fremantle.

In the early 1970s, skindivers became interested in the wreck. It was reported to the Western Australian Museum in 1974 by Joe Castlehowe and John Bell of Albany who were given a reward of $150 by the Trustees of the Museum. The wreck lies on the east side of the entrance to Oyster Harbour in latitude 34° 59.9′ south, longitude 117° 56.8′ east. The figurehead now decorates the Residency Museum at Albany. A large iron pot excavated from the site in 1977 by archaeologists from the Western Australian Museum is undergoing treatment at the Conservation Laboratory in Fremantle. (See map on p. 126.)

The place and date of building of the *Lady Lyttleton* (Official Number 32704) are unknown. The vessel arrived at Sydney in 1861 as the *Sultan*, and when she departed for Newcastle in the same year, had been re-named the *Lady Lyttleton*.[4] She was owned by Harold Smith of Melbourne. In May 1867 he empowered Captain McArthur to sell the ship for a sum not less than £10 at any port out of the Colony of Victoria![5] The *Lady Lyttleton* had one deck, three masts, a square stern, a wooden frame and a bust figurehead. Her dimensions were 28.8 metres by 6.4 metres by 3 metres.

NOTES

1. *Perth Gazette*, 5 July 1867.
2. Trevor Tuckfield, 'Lady Lyttleton in Emu Point Channel', *The Albany Advertiser*, 22 March 1971, p. 3.
3. *Inquirer*, 31 July 1867.
4. Register of Arrivals, Sydney.
5. Register of British Ships, Melbourne.

Divers prepare for lifting a part of the *Lady Lyttleton*'s cargo—vats used for boiling down sheep for soap.

Unidentified whaler, 1867

In July 1867, Police Constable Joseph Watson searched the coast south of the Irwin River because of apprehension about the non-arrival of the coaster *Speculator*. After travelling about 40 kilometres along the beach, he came upon the tracks of a man near a cave. The tracks, showing a nailed boot, continued another 65 kilometres south and then turned inland. While following the tracks, Watson found several masts, yards, a jib-boom, an oar, a small box, an old cask, a large ship's rudder, and various other items.[1] The coaster *Speculator* arrived in port later, but there was conjecture that the wreckage might have come from the lost *Emma*. (See map on p. 44.)

The Colonial Secretary arranged for the cutter *Victoria* to be sent north so that its crew could examine the coast and the Abrolhos for traces of shipwrecks or shipwrecked persons. At about the same time, the cutter *Albert* was sent across to the Abrolhos for the same purpose. Mr Harford, the master of the *Victoria*, later reported:

> We coasted down to an island further to the east [than Hummock Island in the Easter Group] in the same group, and having anchored the cutter, we proceeded to examine the shore in the dinghy, when we found a ship's boom, about 45 feet long, fresh from the water; a water-cask, evidently just come ashore; a lot of cabin fittings of a ship perhaps of 300 tons burthen, fresh broke away, and an oak plank, about 14 or 15 feet long, by about 3½ inches thick and 14 inches wide, which we brought on board, as also some of the bolts and a block, not at all corroded, which we had contrived to cut from the spar.[2]

Harford was of the opinion that a ship, probably an American whaler, had been wrecked between the Beagle Islands and the Southern Abrolhos. It also seems possible, however, that the Abrolhos wreckage came from the *Cochituate*, and that the mainland wreckage was from the *Leander*.

NOTES

1. *Inquirer*, 7 August 1867.
2. *Inquirer*, 9 October 1867.

Don Juan

On 28 September 1867 the Belgian ship *Don Juan* was run on shore on the north bank in harbour at Albany. She was pulled off a short distance, but a heavy squall sent her back on the bank.[1] The vessel was apparently refloated when the weather moderated.

NOTE

1. *Inquirer*, 23 October 1867.

Favourite

The schooner *Favourite* was built at Bunbury in 1856 for licensed victualler John Morgan. The vessel was taken to Fremantle immediately after launching and the *Inquirer* commented:

> She reflects great credit upon her designer and builder, Jackson, of Bunbury, and is generally admired. Her spars are, however, considered to be too lofty for the size of the vessel, a defect which is about to be remedied.[1]

The vessel was used to supply Bunbury farmers with phosphate fertilizers from the north. In this role, the *Favourite* was involved in several accidents. She went ashore at Bunbury in September 1856, but was apparently refloated without great difficulty.[2] However, a more serious casualty was reported in January 1858, by which time she had been sold to John Bateman:

> The *Favourite*, we are sorry to hear, was wrecked at a reef on the Abrolhos on Tuesday, 15th ult. The sails, spars etc were saved, and, as far as could be ascertained, she had received but little damage. She was so high upon the reef the crew could walk around her at low water. The captain [Goss] and some of the crew started in a whale boat on Saturday, 19th ult., and got as far as King's Table Hills, when they were forced to bear up for Port Gregory, at which place they were on the 20th ult., and

> would leave as soon as wind permitted for Fremantle. The *Favourite* has lost her false keel, her main keel is splintered, and some bolts and trenails have been started. Three men were left on the island, who have another boat in their possession, and sufficient stores to last three months. The owner of the *Favourite*, Mr J. Bateman, proceeds to the Abrolhos as soon as the whale boat arrives from Port Gregory.[3]

Bateman succeeded in refloating the *Favourite* and she traded regularly along the west coast until she was wrecked at Port Gregory in 1867.

On 24 November, the *Favourite* was sailing out of the Port through the Gold Digger Passage when she was put 'aback' and driven on the reef, where she quickly started to break up. Captain Lakey tried, unsuccessfully, to get her off and the *Favourite*, described as having been 'an insect ridden old tub', was struck from the Register.[4]

In 1980, skindiver David Totty reported to the Museum that he had found wreckage on the seaward side of the southern tip of the reef north of the Hero Passage. However, this position conflicts with the locality given for the *Favourite* at the time of her sinking. (See map on p. 53.)

The 46-ton wooden schooner *Favourite* (Official Number 40468) carried two masts, a square stern and a scroll head. Her dimensions were 20.3 metres by 5.8 metres by 2.3 metres.[5]

NOTES

1. *Inquirer*, 20 February 1856.
2. *Inquirer*, 3 September 1856.
3. *Inquirer*, 6 January 1858.
4. Helen Summerville, 'Port Gregory', *Early Days*, 6: 8 (1969), p. 78.
5. Register of British Ships, Fremantle.

Ariel

On 23 November 1867, several of the crew for the pearl shell vessel *Ariel* left Fremantle bound for the north coast, on the coaster *Clarence Packet*. The pearling industry was flourishing:

> The Pearl Shell fishery was most prosperous, and if we are to believe report, notwithstanding the number of boats now engaged in it, the returns average a ton of shells per 27 days for every white man employed, but it must be observed that much of this 'beach combing work'—for it is but little more, is done by natives, but even then the gain must be enormous, considering that a ton of shells will readily sell in the colony for £100, giving £25 a week for each white man engaged in the venture—something like the palmy days of the Victorian gold fields, when surface digging gave such splendid returns; but like the gold fields such golden harvests cannot last long, and the simple means and the small boats now so successful in the shallow waters along the coast will soon find that they have gathered in all the harvest within their reach, and the field of deeper waters will require larger craft fitted with proper diving apparatus, the employment of which will in all probability produce equal if not better results.[1]

There were, nevertheless, several more immediate problems for the pearl seekers. The industry relied on the exploitation of resentful Aborigines as a labour supply, and many of the new chums bringing vessels to the north did not understand the dangers of the cyclone season.

A cyclone struck at the Ashburton River on 4 January 1868. At the time, pastoralist Henry Fisher and his men were camped near some fifty bales of wool that were awaiting shipment. The wool was swept into the mangroves and along the beach, and Fisher and his party sought refuge on a sand-hill, which became surrounded by the sea. The waves carried away large masses of sand from the hill, but then, luckily, the tide turned, leaving the men dry but without rations, clothing or saddles.[2]

At that time, the pearling vessel *Ariel* was lying between two reefs off Locker Point, 50 kilometres west of the Ashburton. Edward Corbett, a survivor, later stated:

> Between 7 and 8 o'clock on the morning of 4th inst. there was a heavy surf from S. West, with strong current of about 7 or 8 knots from the same quarter, setting to the northward. There was a bank between the two reefs upon which we anchored. The wind blew very strong. The vessel was thrown on her beam ends, head on to shore. She had two anchors down, and foundered directly she went over. We were then about a mile and a half from the shore. After she went down I found myself clear of the wreck, swimming in the water.[3]

Corbett reached the shore after swimming for three-quarters of an hour. He climbed a hill, but could not make out any wreckage or other survivors:

> I saw a hundred and eighty odd natives about half a mile off, and did not like to wait longer, so I struck a course N.N.E. until I came to a fine creek about a mile from where we were wrecked.[4]

Corbett then swam the Ashburton and walked about 30 kilometres further to Fisher's camp. The master of the *Ariel*, Joseph Barrett, and two crew-men, drowned. (See map on p. 186.)

A 26-ton schooner named *Ariel* was built at Hobart in 1845 for David Laing. This vessel had a standing bowsprit, a scroll figurehead and a pink stern.[5] Her dimensions were 14.3 metres by 3.8 metres by 1.6 metres. The Register states that she was wrecked in 1868. Several other vessels of the same name were registered at Sydney during the 1860s.

NOTES

1. *Perth Gazette*, 3 April 1868.
2. *Inquirer*, 1 April 1868.
3. *Ibid*.
4. *Ibid*. See also *Herald*, 4 April 1858.
5. Register of British Ships, Hobart, Acc. No. CUS.38, State Archives of Tasmania. A pink is a fishing boat with a narrow stern.

Pearl

The pearling boat *Pearl*, commanded by a Mr McKenzie, was damaged in the same cyclone that claimed the *Ariel*. The *Pearl*, when coming in from pearling, grounded on a bank off Butcher's Inlet and was holed.[1] She was sold for £65 and apparently put back into service, as a vessel of that name was pearling in 1869.[2]

NOTES

1. *Herald*, 4 April 1868.
2. *Inquirer*, 8 December 1869.

Emily

The *Emily* was launched at Fremantle on 25 April 1868. Described as 'a very handsome schooner', she was intended for the coasting trade, and her owners advertised that she would run regularly between Fremantle and the Irwin.[1] She set out on her first voyage to that port on 7 May. After a quick return voyage, she sailed again for the Irwin before the end of the month.

With a crew of four, the *Emily* set out on her second return voyage to Fremantle on 13 June, and that night, a heavy gale blew in from the north-west. The vessel was wrecked soon afterwards.

By the beginning of July, people in Fremantle began to fear that the vessel had been lost. Fragments of wreckage found on the beach at North Fremantle were recognized by Robert Wrightson, the vessel's builder, as having been part of the taffrail of the *Emily*.[2] At the end of that month, pieces of wreck were found on Rottnest, leading to speculation that the vessel had gone down there.

The search was narrowed to the area between Wanneroo and Moore River with the finding, at the beginning of August, of ship's timbers, bags of flour, and the body of a man identified as William Bailey, one of the crew. Soon after, the hull was found on the beach just south of a point, later to be named Wreck Point. It appeared that the *Emily* had anchored under the shelter of the reefs, but had had her hawse-pipes wrenched out by the tempestuous weather and was then driven ashore.[3] The wreck was partly embedded in the sand with the head onto the beach. Some of the cargo of copper ore remained in the hull. (See map on p. 22.)

Bailey's body was taken to Fremantle. The other four men were not found.

The 40-ton schooner *Emily* (Official Number 61083) was a wooden two-masted vessel with one deck and an elliptic stern.[4] Her dimensions were 18.4 metres by 4.9 metres by 2.1 metres. She was owned by Richard Harford, a Fremantle boatman.

NOTES

1. *Perth Gazette*, 8 May 1868.
2. *Perth Gazette*, 3 July 1868.
3. *Inquirer*, 12 August 1868.
4. Register of British Ships, Fremantle.

The *Emily* at Wreck Point.

Sea Nymph

The brig *Sea Nymph* was built at Prince Edward Island in 1850. The vessel was sold to Glasgow in 1853, to Geelong in 1854, to Melbourne, to Adelaide, and finally to Sydney in 1866.

In 1868, the vessel was taken to Busselton for a load of timber. She was brought to Fremantle to complete loading, but drifted ashore at St Catherine's Point during a gale on the morning of 18 June.[1] The wreck, exclusive of anchors, chains etc., was auctioned by Samson and purchased for little more than £50 by Edward Newman, a local merchant. (See map on p. 242.)

The 173-ton *Sea Nymph* (Official Number 31960) had one deck, a standing bowsprit, square stern and billet head.[2] Her dimensions were 27 metres by 6.4 metres by 3.7 metres.

NOTES

1. *Inquirer*, 24 June 1868.
2. Register of British Ships, Geelong.

Northumberland

The 1,168 ton ship *Northumberland* left Newcastle on 2 March 1868 under the command of Captain John Humphrey. The vessel carried 1,750 tonnes of coal for the P. & O. Company's coaling depot at Albany. About 480 kilometres out from Albany, the *Northumberland* met with a heavy storm and sprang a leak on 14 June. By the 16th, there was 3 metres of water in the hold, and it was reaching 3.3 metres 3 days later, despite heavy pumping.[1]

Two topsails were blown out and the cargo shifted, causing the vessel to list. The gales continued, washing away the deck houses. West Cape Howe was sighted on the 20th, but by that time there was 4.6 metres of water in the hold. Captain Humphrey stated:

> We shaped our course along the land as we were afraid the ship would go down under our feet, she was nearly unmanageable. The vessel struck about 10 pm of 20th on a reef off Bald Head. The second time she struck

> she broke her rudder and I believe started her stern post. My object in getting in with the land was to save the crew. When we struck the ship was very deep in the water, from the quantity of water in the hold she took a long time to answer her helm when it was put up. We never saw any light until two or three minutes after the ship struck. She canted with her head to seaward, being then quite unmanageable and giving up all hope of saving the ship, I ordered out the boats, the weather was then moderating, but still a heavy sea on. We remained under the lee of the ship until near day light, then started off for the harbour to try and get assistance. At daylight found the ship still afloat, went back again towards her with the intention to board her. On the way we stopped at Breaksea Island, the Light Keeper hoisted the signal of distress up and seeing the Harbour Master coming off, we waited on the Island. We had not been on the Island above an hour when we saw the vessel go down head first at about 9.30 am as near as I can judge she went down in a direct line between Breaksea Island and Cape Vancouver. We were about 3 miles from Bald Head about S.E. when we abandoned the ship . . .[2]

An ordinance of 1864 had established that when any British ship was wrecked off Western Australia a court of inquiry would be held. The court of inquiry into the wrecking of the *Northumberland* found no blame attached to the master, but that the wreck proved the necessity of a lighthouse on Bald Head. The extra light, on Eclipse Island, did not come into operation until July 1926.[3]

The wreck lies in approximately 54 metres of water and has not yet been found. Some timbers and a wooden knee found at Rocky Point could possibly have come ashore from the wreck, but in later times, hulks were sunk in that general vicinity. (See map on p. 126.)

The wooden ship *Northumberland* (Official Number 50568) was built in 1864 and registered at Liverpool.[4] With dimensions of 55.1 metres by 11.3 metres by 7.3 metres, she was built of spruce, birch, hackmatack and oak.[5] The vessel's owners were Davis and Sons, and she had a crew of twenty-two men.

NOTES

1. John Humphrey, Evidence given at the Inquiry into the sinking of the *Northumberland*, 26 June 1868, C.S.R. 621, fol. 55.
2. Humphrey, evidence.
3. *The Albany Advertiser*, 22 February 1971.
4. *Lloyds Shipping Register*, 1868.
5. *Registre Veritas*, 1868.

Albatross

Strong storms during June 1868 played havoc with shipping along the west coast. On the 15th, the coasters *Sea Bird* and *Twinkling Star* dragged ashore at the Irwin River, and the cutter *Albatross* was sent down from Geraldton on the 20th with carpenters and equipment to refloat them. This task accomplished, the *Albatross* sailed on the morning of the 24th from the Irwin for Champion Bay. On this occasion, she had on board a number of Messrs Bateman's whaling party.[1] The *Sea Bird*, which accompanied her, returned to the Irwin the same afternoon, finding the north-westerly winds too difficult to proceed against.

The *Albatross* was unable to weather Point Moore, so she had to turn back. The vessel was about 1.5 kilometres off shore at the mouth of the Irwin River at about 6 p.m. when a heavy sea struck and swamped her.[2] The mast was carried away, together with everything on deck, and all the passengers were thrown into the sea. The *Albatross* rolled over and over in the trough of the sea, until she grounded on the bar at the mouth of the Irwin River. Seven men and a woman were drowned, and three men swam ashore. (See map on p. 44.)

The *Albatross* was a new boat of 18 tons, licensed for harbour work. Her master and owner, the shipbuilder William Garrard, of Geraldton, was drowned in the wreck.[3]

NOTES

1. *Inquirer*, 15 July 1868.
2. Resident Magistrate, Geraldton, to Col. Sec., 29 June 1868, C.S.R. 626, fol. 83.
3. Resident Magistrate, Geraldton to Col. Sec.

Nautilus

The pearling vessel *Nautilus*, owned by a Mr Wardell, was caught in the same cyclone that wrecked the *Ariel* on 4 January 1868. The *Nautilus* was blown clean over the mangroves fringing the coast of Nickol Bay, and was still there 7 weeks later. Another pearl shell vessel, the *Lone Star*, was driven ashore, but was repaired and refloated not long after.[1]

The master of the *Nautilus*, a Mr Jarman, erected a tent near the mangroves. About this time, Aborigines stole some flour from the *Pearl*, a small vessel engaged in pearl shell fishing, and as a result, Police Constable Griffis set out with an Aboriginal assistant named Peter to apprehend those responsible. The two were armed with a revolver, a carbine, a good supply of ammunition and warrants for the arrest of no less than fourteen Aborigines. They caught one man, named Coolyerberri, and put a chain around his neck.

At this time, the Aborigines became alarmed and held a council to decide what they should do. They attacked at night, on 7 February, spearing to death Griffis and Peter, and freeing Coolyerberri. George Breem, a sailor from the wrecked *Nautilus*, had shared his tent with the policeman. He too was killed.[2] Jarman was out collecting shell when the camp was attacked. Later searchers were unable to find him, and it was thought that either he had fled, and then died of exhaustion, or he had been killed.

The Government Resident, Robert Sholl, asked the pastoralist Alexander McRae to organize a posse, empowering him to swear in as special constables every member of the party. A large number of tracks were followed west to the Mermaid Straits. The Aborigines fled on seeing the approaching posse, but some were cut off and shot. More Aborigines were seen on the islands in Flying Foam Passage, and after pursuit, several of them were shot.[3]

A second posse, led by pastoralist John Withnell, took the cutter *Albert* to seek the culprits among the islands. Two prisoners were taken, but they were shot while trying to escape.[4]

It was later estimated that sixty Aboriginal men, women and children were shot dead during the Flying Foam massacre. McRae and Withnell were officially thanked for their services rendered to the white community.

The *Nautilus* was repaired and refloated for the following season, but disappeared in April 1869. On the night of the 3rd, the vessel left a creek about 29 kilometres east of the De Grey River, bound for Port Walcott for rations. On board were two Aborigines, two Europeans (Charles Kulp and James Guillard), and over a tonne of shell. Five weeks later, she had not arrived, and it was presumed that she had foundered at sea.[5] (See map on p. 118.)

Again, it seems likely that the *Nautilus* eventually turned up, because a vessel of that name is mentioned as operating at the Fortescue River in 1875 (see *Lily of the Lake* entry).

Nothing is known about the specifications of the *Nautilus* beyond the comment in the *Inquirer* that 'she was too small and slight a boat for such work'.[6] The vessel had apparently first arrived at Nickol Bay on 22 December 1867. '*Nautilus*' was a popular name for small craft in the 1870s. A 21-ton cutter of that name (Official Number 32523), owned by Robert Towns in the 1860s, was listed in the Sydney Register as having broken up, but that vessel was a paddle-steamer, unlikely to have visited Western Australia.[7] A 45-ton schooner of the same name (Official Number 64768) was registered at Launceston in the 1870s.[8] Another *Nautilus*, a 48-ton schooner 20.4 metres in length, was owned by the whaler and lighterman John Tapper, and registered at Fremantle. That vessel is listed as having been broken up at Cossack in 1885.[9] However, at 20.4 metres, it could hardly have been described as being too small for the industry.

Five graves have been located on Dolphin Island, but the skeletons examined appear to be Asian or European, and so can be presumed not to have been associated with the Flying Foam massacre.[10]

NOTES

1. *Perth Gazette*, 3 April 1868.
2. *Ibid*.
3. 'Report of Mr McRae', *Inquirer*, 4 April 1868.
4. 'Report of Mr Withnell', *Inquirer*, 4 April 1868. See also T. Gara, The Flying Foam Massacre: An Incident on the North-West Frontier, Western Australia, in Moya Smith (ed), *Archaeology at ANZAAS, 1983* (Western Australian Museum, 1983) pp. 86–94.

5. Sholl to Col. Sec., Roebourne, 7 May 1869, C.S.R. 647, fol. 23.
6. *Inquirer*, 4 April 1868.
7. Register of British Ships, Sydney.
8. Register of British Ships, Launceston.
9. Register of British Ships, Fremantle.
10. Jack MacIlroy, Dampier Archipelago Historic Sites Survey, 1979 (Report prepared for the Australian Heritage Commission), p. 36.

Faith

The cargo boat *Faith* was wrecked on the bar at the mouth of the Swan River on 5 August 1869. On the afternoon of 4 August the vessel was laden with a general cargo from the *Hastings*, that was to be taken up river to Perth, but first the passage through the bar had to be negotiated. James Reilly, master of the *Faith*, later stated:

> I arrived at the mouth of the bar at about ½ past 7 o'clock the following morning. It was not blowing hard at that time. What wind there was was from the eastward with a spring tide running in. There was a nasty sea outside, but the passage of the bar was perfectly smooth. The sail was set, and we tried to pole the boat in. We nearly got the boat into the near passage, when a sea struck her, and she became unmanageable from the force of the tide, and went on to the reef on the south side of the channel. Shortly after she became a total wreck.[1]

In finding that no fault was to be attributed to Reilly, John Bateman, appointed Nautical Assessor by the court, criticized the new entrance as being unsafe.[2] (See map on p. 242.)

The *Faith* (Official Number 36542) was an 8-ton cargo vessel rigged as a two-mast dandy. She was built at Fremantle in 1861, and owned by William Barnard and Mark Read. Her dimensions were 10.4 metres by 3.1 metres by 1.1 metres.[3] In addition to lightering work, the vessel did at least one voyage to the Murray River.[4]

NOTES

1. Minutes of Court of Inquiry, 10 August 1869, Lands and Surveys Department Records No. 40/2, Acc, No. 645, Battye Library.
2. *Herald*, 14 August 1869.
3. Register of British Ships, Fremantle.
4. *Inquirer*, 4 March 1868.

Swan

The schooner *Swan* was built at Fremantle in 1865 and used in the coasting trade. In 1867, for example, the vessel was taken on a voyage to Albany and Esperance Bay with general cargo.[1]

In late September 1869, the *Swan* departed Fremantle bound for Port Irwin and Champion Bay under the command of Charles Peterson. On the evening of 2 October, the *Swan* was approaching the Irwin. Peterson saw that he could not reach the Port itself before sunset, so at about 8 p.m., he made for the shelter of the reef. At a critical point, Peterson ordered his helmsman to steer sharply to port, but the helmsman did the reverse.[2] Peterson took over, but before the *Swan* answered to the helm a large swell took the vessel onto the reef.

The next morning, pastoralist and businessman William Dalgety Moore, the *Swan*'s owner, rode down to the beach and found the vessel among the reefs. He and the crew unloaded the *Swan*, and the following day, using buckets, lowered the water in the hold by about 1.2 metres.[3] An anchor was run astern to kedge the vessel back several metres. Then the sea rose, throwing the vessel broadside and snapping the cable. The salvage boat struck the side of the *Swan* and was damaged, forcing the men to retire to shore.

The following day the sea was running right over the *Swan*, so Moore decided that, rather than attempting to refloat her, he would concentrate upon stripping the vessel. By the end of the month, the *Swan* had been completely broken up.[4] (See map on p. 44.)

The 24-ton *Swan* (Official Number 36554) had two masts, one deck and a square stern. Her dimensions were 17 metres by 4.3 metres by 1.7 metres.[5]

NOTES

1. Albany Shipping List, 3 December 1867, C.S.R. 606, fol. 92.
2. *Inquirer*, 20 October 1869.
3. *Ibid.*
4. *Inquirer*, 27 October 1869.
5. Register of British Ships, Fremantle.

Flying Scud

Resident Magistrate Robert Sholl noted the loss of the *Flying Scud*, a boat belonging to George King, at a passage in the reef 7.2 kilometres west by north of the *New Perseverance* wreck, on 19 August 1870.[1] After striking the reef, between Point Samson and Cape Lambert, the crew stayed with the vessel for a time, then made for the shore. Boorancabba, an Aborigine, reached shore first and spread the alarm. Another crewman, J. Carroll, also swam safely ashore, but George King, who was sick, drowned in the attempt.[2] (See map on p. 118.)

In November some of the goods from the wreck were salvaged by D. Chapman.

NOTES

1. R. J. Sholl, Occurrence Book, Acc. 194, Battye Library.
2. *Inquirer*, 21 September 1870.

Champion and Sustenance

Three vessels, named *Champion*, *Sustenance* and *Kate Mullet*, sailed from Port Walcott for Depuch Island some days before the cyclone of Christmas Day 1870.[1] After the cyclone, it was feared for some time that all three had been lost. In March, however, it was reported that, although the *Champion* was a complete wreck, the *Kate Mullet* had gone farther to the eastward and was safe.[2] Nothing was said of the *Sustenance*, and it is not known whether or not that vessel was relocated. (See map on p. 113.)

The *Mercantile Navy List* records a 73-ton *Champion* (Official Number 31880) registered at Sydney, and the Register of British Ships at Sydney lists a 5-ton cutter of that name, supposed to have been lost in 1868.[3] No details of the *Sustenance* are available.

NOTES

1. *Inquirer*, 15 February 1871.
2. *Inquirer*, 8 March 1871.
3. *Mercantile Navy List*, 1861. Register of British Ships, Sydney.

Melanie

The 133-ton three-masted schooner *Melanie* (Official Number 32356) was owned in Sydney by the prominent shipowners Messrs R. Towns and Co., and a Captain Edward. The vessel was engaged for several years in the bêche-de-mer trade in the Torres Straits. She was fitted out as a whaler, hoisting three boats, which made her particularly suitable as a roving pearler.[1]

In 1869, the *Melanie*, in company with another Sydney vessel named *Kate Kearney*, sailed around the north of Australia via Torres Straits to Roebourne, where the crew related one of their adventures to a visitor from Fremantle:

> While the *Melanie* was lying in Somerset (Cape York), Captain McEnroe went over to Prince of Wales Island, some 40 miles distant, to trade for tortoise shell, the natives in the canoes who boarded the vessel said they

> had none with them, but plenty at the camp, and evinced great unwillingness to take the party to where it was, the captain insisted, and on arriving at the camp he found several articles of clothing, pieces of sails, rigging etc, that convinced him that some vessel had been wrecked or taken, he immediately seized some of the natives, and they then told him that some men of another tribe had taken another vessel some days before, killed the crew, and scuttled and burnt her. He returned to Cape York and told Mr Frank Jardine, the police magistrate, what he had seen, and that gentleman being a Queensland magistrate, and consequently not hampered by any absurd aboriginal protection instructions, immediately organised a party of mainland natives (Zardigans) and assisted by some Tanna men from the *Melanie* went over to the island, crept on the dastardly wretches at daylight, and 'dispersed' the men of the tribe, justly avenging a most brutal and unprovoked outrage.[2]

The vessel taken by the Islanders was the 16-ton, Batavia-registered cutter *Spurwar*, commanded by Captain Gascoyne, who had earlier visited Nickol Bay in the *Jeannie Oswald*. Gascoyne was pearling in the Straits with a Malay crew when he was killed.

The *Melanie* carried fifty Solomon Islanders and ten Europeans. Upon arrival in the West, the *Melanie*'s captain spread his crew, and local Aboriginal recruits, along the coast in separate parties from Roebuck Bay to Roebourne.[3]

On Christmas Day, 1870, a cyclone swept the coast. At Depuch Island, a vessel named *Melanie* was reported wrecked with no loss of life.[4] The questions arise as to whether the reports referred to Towns' 133-ton schooner or a smaller vessel, and whether the 'wrecked' vessel was subsequently refloated. Towns' schooner was sold to the Port of Dunedin in New Zealand in 1873, so it is clear that if his vessel sank during the 1870 cyclone, it was subsequently refloated.[5] (See map on p. 32.)

The wording of the newspaper reports seems to indicate that it was another, smaller, *Melanie* that was lost at Depuch Island. The *Inquirer* referred to the wrecked *Melanie* as a 'shell boat' and as an 'open boat'.[6] Although, in normal circumstances, it would be surprising if a 133-ton schooner was referred to as a 'boat' rather than a 'ship', it is nevertheless true that it became common practice for vessels employed in the pearling industry to be referred to as 'boats', irrespective of size and equipment. However, the term 'open boat' does not correspond with the 133-ton schooner. If it was a smaller, open boat that was sunk, it may never have been registered.

NOTES

1. *Herald* , 30 October 1869.
2. Richard Thatcher, 'The Pearl Station on the North West Coast', *Herald*, 30 October 1869.
3. *Inquirer*, 15 September 1869. See also *Inquirer*, 5 January 1870.
4. *Inquirer*, 15 February 1871.
5. Register of British Ships, Sydney.
6. *Inquirer*, 15 February 1871.

Coquette

The *Coquette*, like the *Melanie* and the *Kate Kearney*, was a Sydney-registered schooner that became involved in the pearl shell fishery in the Torres Straits in the 1860s. In 1868, the 72-ton *Coquette* (Official Number 32638) was sailed around the north coast with the *Kate Kearney*, and the two vessels, both owned by a Captain Barnes, were used for collecting bêche-de-mer and pearl shell off the North-West. In August of that year, they were operating in the Flying Foam Passage.[1] Between them, the South Sea Islander crews of the two vessels collected about 17.5 tonnes of shell and a tumblerful of pearls. These were sent to London for sale.

The *Coquette* had another season of shelling off the North West in 1869, this time in company with the *Melanie*. When the cyclone struck on Christmas Day, 1870, a vessel named *Coquette* was reported wrecked at Ricoe Bay, in Nickol Bay, without loss of life. (See map on p. 118.)

Similar questions can be asked about the *Melanie* and the *Coquette*. The 72-ton schooner *Coquette* was eventually totally wrecked at Clarence River

Heads in 1872.[2] Like the *Melanie*, the *Coquette* was described in the *Inquirer* as an 'open boat', and it seems likely that it was a small, unregistered vessel named *Coquette* that was wrecked in Nickol Bay.[3]

NOTES

1. *Perth Gazette* , 4 September 1868.
2. Register of British Ships, Sydney.
3. *Inquirer*, 15 February 1871.

Pilot

The cutter *Pilot* was built on the Swan River at Perth in 1858 by John Mews, her original owner. In 1868, under the command of her then owner Captain Charles Adams, the vessel was taken, in ballast, to the North West to be used for pearling.[1] A season's progress report in December 1869 noted that the *Pilot* and the *Pearl* had, between them, brought in 3 tonnes of shell.[2]

In December 1870, the *Pilot* was at Butcher's Inlet in company with the cutter *Mystery*, the schooner *Mary Ann*, and the pearling boats *Crest of the Wave* and *Bonnie Dundee*. On Christmas Day, the cyclone struck. The *Mystery* was driven above high water mark, but sustained no damage. On board the *Mary Ann*, the barometer fell to 714 millimetres, lower than had ever been recorded there before, but the *Mary Ann* did not go ashore.[3] The *Crest of the Wave* and the *Bonnie Dundee* both foundered. The cutter *Pilot* was blown ashore on the west side of the inlet, her rudder and four planks broken and all her stores damaged. In the settlement, ten houses were completely blown down and the other few were more or less damaged. (See map on p. 118.)

In March, it was reported that the *Mystery* had been refloated and had departed for the De Grey River on 19 January, while the *Bonnie Dundee* had been repaired and had started pearling. But the *Pilot* was lying in the Inlet with a hole in her bottom. It must either have been repaired or had the appearance of being capable of being repaired, because Resident Magistrate Robert Sholl's sons Robert and Horace bought it in September 1872. The *Inquirer* was probably referring to another accident to the same boat in 1873:

> The cutter *Pilot* had foundered in the Creek in consequence of several of her planks having started. This affords a proof of her unseaworthiness, and suggests the necessity of the early appointment of an official to see that the numerous coasting craft are fit to proceed to sea, and that, moreover, they are properly provisioned and in the charge of competent seamen.[4]

The Register notes that there had been no trace of the vessel since 1872, although the registry was not closed until 1908.[5] The *Pilot* (Official Number 36537) was an 8-ton vessel with the dimensions 9.5 metres by 3.3 metres by 1.2 metres. She was carvel built with a square stern. The Fremantle shipwright Robert Wrightson, espousing the virtues of jarrah as a boatbuilding timber, said that, in 1866, he repaired a small 18-year-old lighter called the *Pilot*, which had been stove in by the weight of a large steam boiler falling into her. Wrightson remarked that the original jarrah planking and frame timbers were perfectly sound.[6] If Wrightson's estimate of the age of the vessel is ignored, then it could be the same craft.

NOTES

1. *Inquirer*, 26 February 1868.
2. *Inquirer*, 8 December 1869.
3. *Inquirer*, 15 February 1871.
4. *Inquirer*, 11 June 1873.
5. Register of British Ships, Fremantle.
6. Robert Wrightson to Clerk of Works, 26 November 1870, in *Votes and Proceedings of the Legislative Council, 1871* (Govt. Printer, Perth), p. 79.

Crest of the Wave

The pearling vessel *Crest of the Wave* was lying at the lower landing, inside Butcher's Inlet when the cyclone of Christmas Day 1870 struck. The vessel was carried out to sea and disappeared. While the storm raged, the two crewmen could be heard calling, but nothing could be done for them. George Blurton, one of the crew of the cutter *Mystery*, wrote:

> At 1.25 a.m. on the 25th, at the height of the gale, I heard Howard [who was on board the *Crest of the Wave*] shouting—I believe he said we are sinking. There was only a small boat belonging to Mr Grant on the beach at the time; we could not launch her, as there was a very heavy sea, and it was blowing so strong that we could not stand. Howard was shouting about ten minutes, and then all was still; the tide was on the ebb at the time; my mate J. Johnston sung out to them to slip and run her ashore. I am sure they could not hear him on account of the wind. We did not know she was gone till daybreak, when I found a bag with clothes belonging to Howard; it was right opposite to where the boat was moored; we also found some of her hatches on the beach where she was anchored; my natives found one of her anchors where she lay; also two quarter casks, that I lent them, which I know were in her hold, and one she carried on deck; my opinion is she went down at her anchor, and the ebb tide took her out to sea; the men Howard and Percival I know to my certain knowledge could not swim, the tide must have taken them out with it.[1]

Further down the Inlet, others found the *Crest of the Wave*'s tiller arm, hatch covers and other minor flotsam. The hull itself was not found after the cyclone and must have been taken right out to sea after sinking at its anchors. No details of the *Crest of the Wave* have been located. (See map on p. 118.)

NOTE

1. *Inquirer*, 15 February 1871.

Fitzroy

The 572-ton barque *Fitzroy*, a regular trader from London, struck on the Stragglers Rocks early on Monday morning, 20 March 1871. Captain James Maillard had tired of waiting for the pilot and decided to bring the vessel in himself. The *Fitzroy* was on a port tack, running from Rottnest to Fremantle, when she struck a bank. Maillard then tacked to starboard and ran onto the Stragglers.[1] The steamboat *Lady Stirling* was used to tow the *Fitzroy* off the Rocks and in to the South Jetty at Fremantle, where her cargo was discharged.

The *Fitzroy* was later taken to Careening Bay for repairs. Captain Maillard was found guilty of gross misconduct by a court of inquiry, but as the Nautical Assessor did not concur with the view of the court, Maillard's certificate was not suspended.[2]

NOTES

1. *Inquirer*, 22 March 1871.
2. *Inquirer*, 26 April 1871.

Blue Jacket

In December 1871, a figurehead was found washed up on Rottnest Island. The Port Pilot, George Forsyth, examined the item:

> The wreck consists of the figurehead of a large sized ship and represents the body and shoulders of a man, the head, neck and back part of the figure being evidently destroyed by fire, the figure is dressed in a loose open coat with yellow buttons, the coat appears to have been blue from the appearance of the paint left in the crevices, there is no waistcoat but a loose shirt and large knotted neckerchief round the neck, with the ends flowing loosely over the chest, it has also a broad belt round the waist with a large square buckle on it, and what appears to be the kilt of a cutlass on the left side. From the waist downwards is all carved work with

> two scrolls running from the belt down and apparently ending in the stem, the following words are cut in the scroll and appear to be gilded; on the starboard side is 'Sharp Look Out' and on the port side is 'Keep a Sharp Lo . . .', the remainder of the scroll being away. The figure is cut out of pine or soft wood and is very much chafed, I should say by grinding and rubbing over the reefs it has evidently not been long in the water, from there being neither barnacles or seaweed on it, and no worm holes in it, and the charred wood at the back appears quite fresh. There are five large iron bolts in it, of 2 inch iron, about 5 feet long which have connected it with the stem, the bolts are perfectly straight and the wood appears to have burnt away from them. The figure itself is about 6 feet in length and I should say has belonged to a vessel of at least 1,400 tons register.[1]

Forsyth was right about the tonnage. It later became evident that the figurehead had come from the 1,790-ton clipper ship *Blue Jacket*, a well-known vessel that, in 1855, made the magnificent run out from England to Melbourne of 69 days. The *Blue Jacket* sailed from Lyttelton to London on 1 February 1869 with a cargo of wool, hides and tallow, and a large quantity of gold. She caught fire and was abandoned off the Falkland Islands on 9 March 1869. The figurehead had been brought to the Australian coast by the westerly winds and currents prevailing in high southern latitudes. It was taken to a store near the corner of Cliff and High Streets in Fremantle owned by Captain Owston, at the time Lloyds surveyor in Fremantle, who, it is believed, later took the figurehead to England.[2]

NOTES

1. G. Forsyth to Col. Sec., 7 December 1871, C.S.R. 702, fol. 204.
2. *West Australian*, 14 December 1933.

Strathmore

The 450-ton barque *Strathmore*, a regular London to Fremantle trader owned by the well-known firm of Wilson and Company, struck the Western Australian coast in the early 1870s. Alexander Findlay mentions her in the 1876 edition of his *Directory for the Navigation of the Indian Ocean*:

> Point Cloates lies about 42 miles southward of North West Cape, and off the point, about 12 miles to the westward a vessel named the *Strathmore* struck a rock, having about 8 ft of water over it. Position approximately given 27° 37′ S, long. 113° 24′ E.[1]

However, the vessel was not wrecked. She sank off Birkenhead, while returning to Liverpool via Singapore, as the result of a collision with the *Glengaleer* at anchor.[2]

NOTES

1. Alexander Findlay, *A Directory for the Navigation of the Indian Ocean* (3rd edition, London, 1876), Vol. 1, p. 327. We are grateful to A. C. F. David, of Taunton for alerting us to this passage. Findlay's given latitude fits the Abrolhos rather than Point Cloates.
2. Board of Trade Wreck Register, BOT/19, fol. 243, National Maritime Museum, England.

Adur

The schooner *Adur* struck a rock near the De Grey River at the beginning of 1872. The vessel was beached and found to be badly damaged, but it was reported that she was being repaired.[1] The schooner indeed changed hands several times after the 1872 stranding. A vessel of that name was pearling east of Tien Tsin harbour in December 1875.

The Register was not closed until 1908, although there had been no transactions since 1874, when the vessel was bought by Messrs Grant, Harper and Anderson, of the De Grey River. The 25-ton two-masted schooner *Adur* (Official Number 61093) was built at Perth on the Swan River in 1870, and was owned at that time by Gabriel Adams and others. It seems likely that the pioneer, squatter, merchant and pearler Charles Broadhurst was involved in her operation at an early date. In November 1872, Broadhurst wrote about a Fremantle diver named Edward Thomson who had been working from the *Adur*, so the vessel may have been one of the earliest in the Western Australian pearling industry to have had breathing apparatus used on board.[2]

The carvel-built *Adur* had one deck, a round stern and the dimensions 16.3 metres by 4 metres by 2.3 metres.[3]

NOTES

1. *Inquirer*, 7 February 1872.
2. Charles Broadhurst to Col. Sec., 20 November 1872, C.S.R. 727, fol. 233.
3. Register of British Ships, Fremantle.

Unnamed Lighter and Unnamed Boat at Geraldton, 1872

In what was described as a cyclone at Geraldton in February 1872, a lighter (the only substantial vessel in the bay) foundered at her anchors off the jetty, and a boat that had been beached near the Chapman River was stove in and rendered useless.[1] No further details are available. (See map on p. 53.)

NOTE

1. *Inquirer*, 21 February 1872.

Champion Bay in 1869.

Argo and Dawn

The schooners *Argo* and *Dawn* were driven on the reefs at the Irwin on the morning of 10 March 1872. Both vessels were refloated, the *Dawn* being back in Fremantle by 5 April, and the *Argo* by 15 April.[1] On arrival, neither vessel was reported to have been badly damaged. The *Argo* (Official Number 61081) was broken up in 1889 after years of whaling, coasting work and pearling.[2] The *Dawn* (Official Number 61094) was finally wrecked on a voyage from Broome to Fremantle in 1902.[3]

NOTES

1. *Inquirer*, 10 April 1872 and 17 April 1872.
2. Register of British Ships, Fremantle.
3. Register of British Ships, Fremantle.

Vasse and Bunbury.

THE fine clipper Schooner "DAWN," 60 tons, Robert Owen, master, will sail for the above ports on or about the 3rd of February. This vessel has superior accomodation for passengers; and all goods intended for shipment will be stored free of charge.

For freight or passage apply to
T. & H. CARTER & Co.
Fremantle, Jan. 24, 1870.

An illustrated advertisement from the *Perth Gazette*, 2 February 1870. The vignette is not that of the *Dawn*, but a standard image.

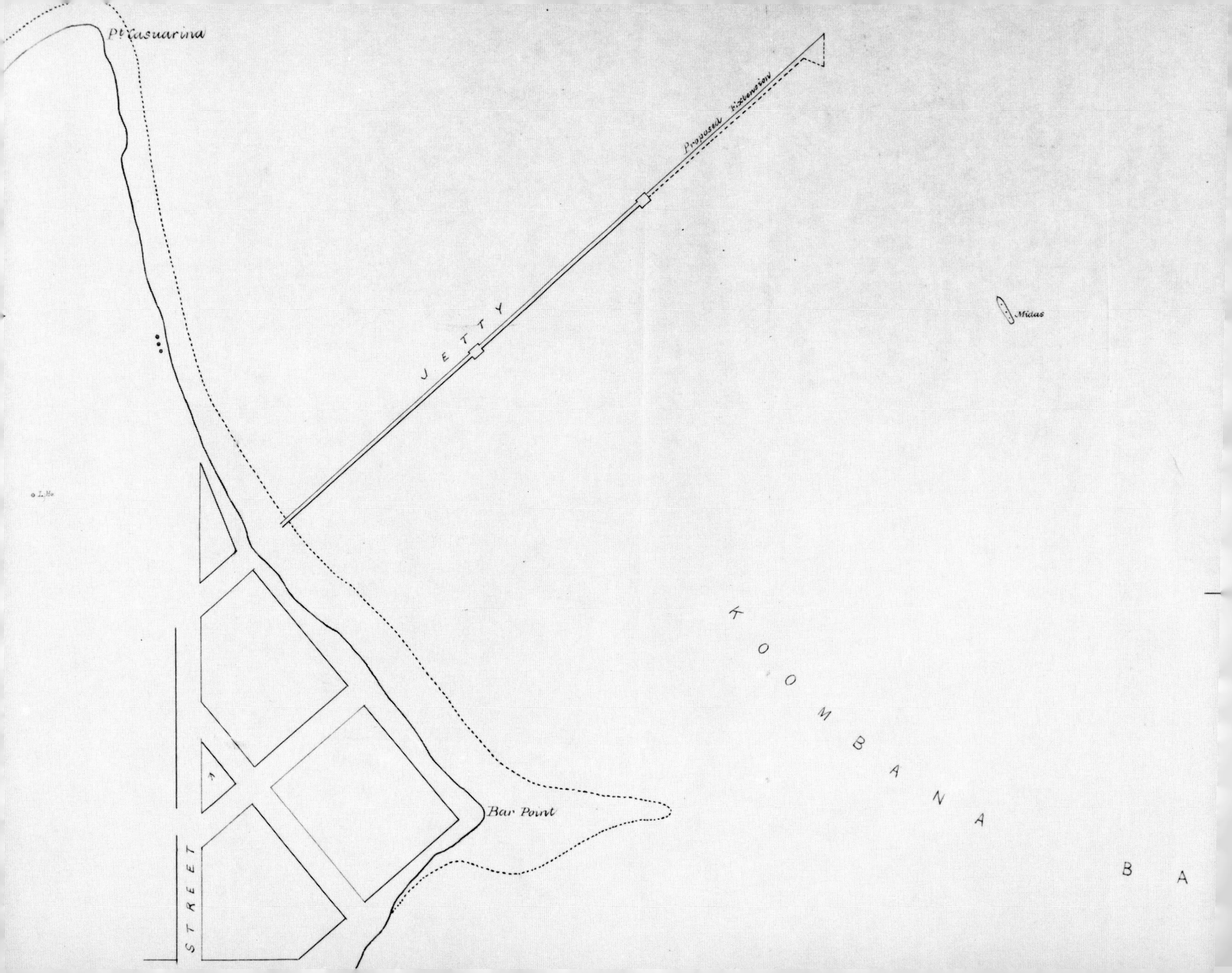
Pt Casuarina
Proposed Extension
JETTY
Midas
KOOMBANA BA
Bar Point
STREET

Midas

The barque *Midas* sailed from Dunedin, New Zealand, on 2 October 1871 for Western Australia. The vessel arrived at Bunbury on 3 February 1872 and commenced loading timber for Lyttelton. Five weeks later, the loading had been completed, the cargo consisting of about 100 twenty-one-metre piles, about 120 other piles and a quantity of sawn timber—in all about 450–500 loads.[1]

When the anchor was dropped, the wind had been blowing east-south-east off the land. But on Sunday, 10 March 1872, the wind suddenly shifted to north-north-east with a violent squall. According to observers, the sea immediately rose, with the appearance of a tidal wave coming into the bay.[2] The crew of the *Midas* dropped a second anchor as the seas began to break over the bows and fly from stem to stern. The *Midas* dragged her anchors some 450 metres and began striking heavily on the bottom. The rudder was carried away and the mainmast broke off close to the deck. The mizzen-mast was chopped down to ease the strain on the vessel. John Cumming, master of the *Midas*, stated at the inquiry:

> At three o'clock I found the vessel to all appearance bilged on the Port side: and by four o'clock the water up to the after hatch way and she was fairly settled down on the ground fore and aft, the sea still breaking right over her and expecting the Poop bulk heads every moment to be stove in. By six o'clock the wind had veered to the westward and moderated and the sea went down.[3]

Two smaller vessels were in the port at the same time, and both parted with their anchors. As a last resort, their masters headed for the bar into the river, which they succeeded in crossing safely. The ketch *Wild Wave* lost her jib-boom, anchors and chains, while the cutter *Twilight* lost her rudder and anchors.

After spending a night on his stranded vessel, Captain Cumming went ashore and asked that the *Midas* be surveyed. The barque was condemned as a wreck and sold, the ship for £120 and the cargo for the same amount. (See map on p. 204.)

Many years were to pass before the wreck was forgotten. In August 1872,

› *Midas* wreck at Bunbury.

the Harbour Master began to express concern about the continued presence of the wreck in his harbour. The purchasers of the hull removed the anchors and the standing rigging, with the exception of the foremast. But, as the deck held firm, they were unable to get at the timber piles inside the hold. In March 1873, the owners of the wreck were hoping that the first winter storm would break it up and scatter the piles on the shore.[4] The authorities, however, were alarmed at the prospect of the damage that might be done to other ships by floating piles. A few of the piles were removed that winter, but the bulk of the wreck, and its cargo, continued to present a navigational hazard.

In March 1874, the Government called for tenders for the removal of the wreck. The sum of £487 was paid to Mr W. F. Stevens to do this task.[5] Stevens used explosives to break up the wreck, and recovered over 200 sleepers, twenty-seven piles, forty iron knees, about 0.25 tonnes of copper, and other assorted items. It was reported that the wreck had been completely cleared away. Yet in December 1874, Fremantle Harbour Master, George Forsyth, inspecting the site for the Government, found there to be only 1.4 metres of water over the site at high tide in an area that was otherwise 3.6–4.3 metres deep. Forsyth wrote:

> . . . from what I saw and felt I feel confident the major portion of the bottom of the ship is still laying there, and a diver will have to be employed before it can be thoroughly cleared away. As it stands at present it is in the highest degree dangerous to small and large vessels using Bunbury.[6]

In May 1875, it was reported that the wreck lay about 365 metres east-south-east of a bend in the jetty with her head to the north. The architect and builder James Manning suggested that a diver use dynamite in rubber-lined bags for blasting underwater. But the Colonial Secretary felt that by delaying until the following summer, the job could be done more easily. In August 1876, the correspondent for the *Western Australian Times* wrote:

> In walking down the jetty on a clear day, if you cast your eye over where she went ashore, you can see the dark patch where she still lies.[7]

Improvements have substantially altered the harbour since the 1870s, and the position of the wreck today is not known.

The 555-ton barque *Midas* (Official Number 33205) was owned by Charles Clark, of Dunedin.[8] It was built in 1865 at Farmingdale, in the United States, of oak, hackmatack and pine.[9]

NOTES

1. Minutes of Court of Inquiry regarding the Stranding of the Barque *Midas* on 10 March 1872, C.S.R. 727, fol. 30.
2. *Midas* Inquiry.
3. *Midas* Inquiry.
4. Resident Magistrate, Bunbury to Col. Sec., 5 March 1873, C.S.R. 813, fol. 125
5. *Herald*, 28 March 1874.
6. G. Forsyth to Col. Sec., 2 December 1874, C.S.R. 813, fol. 153.
7. *Western Australian Times*, 11 August 1876.
8. *Mercantile Navy List*, 1872. See also Board of Trade Wreck Register, 1872, National Maritime Museum, England.
9. *Registre Veritas*, 1870.

George H. Peake, Rio, and Laughing Wave

During a gale at Fremantle on 10 March 1872, three vessels were stranded. The brig *George H. Peake* parted her anchor chain and drove ashore on a bank near Woodman's Point. The brig *Rio* followed onto the same bank. The brig *Laughing Wave* broke her cable and fouled the barque *Sea Ripple*, carrying away that vessel's jib-boom and her own bowsprit, jib-boom, top-gallant mast and part of the bulwarks. After clearing herself, the *Laughing Wave* drifted towards shore and, as she bumped along, parted with her rudder before coming to rest to the north of Woodman's Point.[1]

Each of the vessels was refloated, and a court of inquiry decided that no blame was attributable to any person.[2]

NOTES

1. *Inquirer*, 13 March 1872.
2. Minutes of Inquiry into Casualties that happened to Shipping on 10 March 1872, C.S.R. 727, fol. 22.

May

The 25-ton cutter *May* was stranded at Geraldton on the morning of 10 March 1872. It had been blowing very heavily during the night and the vessel began to drag her two anchors at about 2.30 a.m. The vessel went up on the beach, but was not damaged. When the unusually high tide subsided, the vessel lay 25 metres from the water's edge. The master of the *May*, Vincent, had refloated the vessel by the time of the court of inquiry, which exonerated him from any blame.[1]

NOTE

1. Minutes of Investigation into Cause of Stranding of the Cutter *May* on 9 (*sic*) March 1872, C.S.R. 727, fol. 52.

Flying Foam

The *Flying Foam*, a schooner of 33 tons, was built at Fremantle in 1861 by William Jackson. The *Inquirer* described her in glowing terms at the time of her launching:

> She is a beautiful model, and it is supposed will sail remarkably fast. Her cabin accommodation is far superior to that of any coaster, and will be a great boon to passengers between Fremantle and Champion Bay.[1]

The *Flying Foam* was lost after a decade of coasting work. The schooner left Champion Bay on 6 March 1872, with three passengers bound for Fremantle. Before embarking upon the voyage to Champion Bay, she had damaged her stern by coming into contact with the jetty at Fremantle. The damage was patched over with canvas![2]

The cutter *May*, which arrived at Champion Bay on 8 March with the help of a breeze from the south-east, reported having seen the *Flying Foam* several kilometres west of Champion Bay, shaping a course for Fremantle. The wind continued from the east the following day, but turned into a gale from the north-west on Sunday, 10 March.

Captain John Vincent, of the *May* , thought it was likely that the *Flying Foam* had been driven onto the Abrolhos, so a group of Geraldton residents chartered the schooner *Clarence Packet* to search the south-west islands of the Group.[3] Nothing was found on the Abrolhos or the mainland to indicate where the vessel had gone down. It seems likely that the *Flying Foam* foundered in deep water south of the Abrolhos during the gale. (See map on p. 53.)

The *Flying Foam* (Official Number 36544) was owned by John Bateman of Fremantle. She had one deck, two masts and a square stern. Her dimensions were 18.3 metres by 4.8 metres by 2.1 metres.[4]

NOTES

1. *Inquirer*, 25 September 1861.
2. George Elliot to W. Clifton (Collector of Customs), 13 April 1872, C.S.R. 727, fol. 59.
3. P.C. James Thomson to W. Piesse (Resident Magistrate, Geraldton), 13 April 1872, C.S.R. 727, fol. 55.
4. Register of British Ships, Fremantle.

Conch, Bonnie Dundee, Square and Compass and Two Unnamed Pearling Boats

Another disastrous cyclone struck at Roebourne on 20 March 1872, blowing down all the public buildings, two hotels, several stores and a number of residences. Six pearling vessels were moored in Butcher's Inlet. The *Conch*, a 5-ton vessel, was swept westward 3 kilometres into a large marsh, where she lay badly damaged. The *Bonnie Dundee* (which had foundered during the 1871 cyclone, but was recovered and repaired) was sent ashore a kilometre from her anchorage, this time irreparably damaged. The *Square and Compass* , owned by James Herbert, had sailed up from Fremantle in October 1869, completing the voyage in less than a week.[1] She was, according to one source, borne some 10 kilometres inland and deposited, much damaged, in a marsh at the foot of a verge of hills.[2] Another source states that she was sent 3 kilometres in the direction of the Upper Landing and was irreparably damaged.[3] A boat belonging to stockowner and pearler David McKay disappeared and was not found, and a 3-ton boat belonging to Edward Chapman was smashed to pieces against some stones at the base of a hill about a kilometre from the creek. Another boat, the *Maggie*, was recovered without much damage. (See map on p. 108.)

No other details of the vessels are available.

NOTES

1. *Inquirer*, 13 October 1869.
2. *Herald*, 18 May 1872.
3. *Inquirer*, 15 May 1872.

Pearling Boat at Port Hedland

The *Herald*, reporting the Roebourne cyclone of 20 March 1872, added:

> A small boat, without any hands as far as I know, on board, was also wrecked at Port Hedland.[1]

No other details are available. (See map on p. 32.)

NOTE

1. *Herald*, 18 May 1872.

CUTTER FOR SALE.

A WELL-BUILT CUTTER, of about 16 or 18 tons, well adapted for pearling on the North-West Coast. She is still on the stocks, and can be seen at any time. If necessary, she could be completed and launched within a month.

GABRIEL ADAMS.

Perth, 1st April, 1871.

Cutter for Sale.

A NEW first-class clipper-built CUTTER, of about 5 or 6 tons, just launched from the yard of the undersigned. She is rigged and in guaranteed order, and is well adapted for a pearling voyage. Terms easy.

LAWRENCE & SON,
SHIPWRIGHTS AND BOATBUILDERS.

April 10, 1871.

Illustrated advertisements for pearling cutters, *Inquirer*, 12 April 1871.

Nellie

On the evening of 19 March 1872, the pearling cutter *Nellie* (or *Nelly*) was anchored in Withnell Bay, on the Burrup Peninsula, near Flying Foam Passage. There were a good many pearl shells packed loose in the vessel. Twelve or fourteen Aboriginal divers, fearing an approaching gale, went ashore to sleep. The *Nellie* had one small anchor down with a lot of chain cable. During the night, the cyclone struck and the *Nellie* either sank at her moorings or drifted out to sea and foundered.[1] The two men who remained on board, Peter Lynch and Joseph Henry, were drowned. A search of the islands was undertaken by the *Minny*, without finding any trace of the vessel.

The wreck reappeared in May 1872, when Captain Cadell reported that he had seen it lying in about 10 metres of water in the western entrance of the Boat Passage. The boom appeared above water and the mainsail was furled on the boom, indicating that the *Nellie* had gone down at her anchor. Captain Cadell's European diver attempted to reach the wreck, but his weights were not sufficient to sink him in the current.[2] Cadell and fellow pearler Edward Chapman had hopes of raising the wreck, but the available records do not show whether they were successful. (See map on p. 118.)

No details of the *Nellie* have been found. The vessel was owned by George Howlett, and had been pearling in the North-West since 1869 or earlier.[3]

NOTES

1. *Herald*, 18 May 1872.
2. Sholl to Col. Sec., Roebourne, 29 May 1872, C.S.R. 714, fol. 90.
3. *Inquirer*, 27 October 1869, Supplement.

Unnamed Whale Boat

More deaths occurred in the pearling industry in June 1872. A whale boat belonging to the river navigator, entrepreneur and pearler Francis Cadell was returning from Condon Creek to Cossack when she was caught in a squall off the Tab-a-Tab Reef, capsized and was lost with all hands.[1] Six Malays were on board. No further details are available. (See map on p. 118.)

NOTE

1. *Herald*, 29 June 1872.

Alexandra and Arabian

The schooner *Alexandra* (Official Number 48420) was built in Maine in 1862. In 1872, she was chartered by the merchants Messrs Crowther and Scott, of Geraldton, to take sandalwood to Singapore. The vessel arrived from Port Darwin and had begun loading when gales struck on 15 June and lasted until 23 June.

On the 16th, the steamer *Xantho* dragged her anchors and barely managed to keep off the beach, although her engine was operating under full pressure of steam. On the 21st, the 24-ton cutter *Arabian* parted her cables and went on shore a little east of the jetty. The *Alexandra* followed at midnight, having parted both of her chain cables.[1]

At the time, it was thought that the *Alexandra* had broken her back. She was described in the press as a total wreck and was advertised for sale.[2] The registry was cancelled. However, in March 1873, the *Alexandra*'s owner engaged Captain Thistleton, by then late of the *Xantho*, in the expectation of refloating her. Soon afterwards, it was announced:

> On Thursday morning the 20th the good ship *Alexandra*, with fore sail, jib, and try sail set, sailed gallantly off the land to which she had so pertinaciously adhered since last June.[3]

The *Alexandra* was to be given the new Official Number 61112, but a letter from the Registrar General in London restored her original number. The vessel arrived in Fremantle on 30 September and, after repairs, resumed trading with ports such as Adelaide, Singapore and Melbourne.[4] In 1875, the *Alexandra* was finally wrecked in North Formosa.[5]

During her stay on the beach at Geraldton, the *Alexandra* changed rig from schooner to brigantine, and increased her tonnage from 234 to 287.[6] Her dimensions were 35.5 metres by 8.1 metres by 5.1 metres.

Nothing further was mentioned in the press at that time about the *Arabian* (Official Number 40478). This vessel, curiously described in the Register as a cutter with two masts and a dandy rig, was not removed from that Register until 1908, and may have been refloated in 1872.[7] The *Arabian*, owned at one time by timber merchant Henry Yelverton, was built in 1857, with the dimensions 14.4 metres by 4.3 metres by 2 metres. A cutter of that name was pearling east of Tien Tsin harbour in December 1875.[8]

NOTES

1. *Inquirer*, 3 July 1872.
2. *Inquirer*, 17 July 1872.
3. *Inquirer*, 2 April 1873.
4. *Inquirer*, 8 October 1873.
5. Board of Trade Wreck Register, 1875, National Maritime Museum, England.
6. Register of Arrivals at Fremantle.
7. Register of British Ships, Fremantle.
8. *Herald*, 29 January 1876.

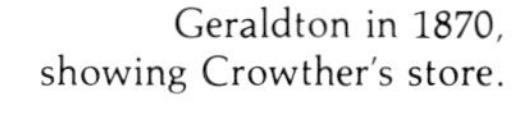

Geraldton in 1870, showing Crowther's store.

Nightingale

The brigantine *Nightingale*, while bringing the first steam locomotive to Fremantle for the Rockingham to Jarrahdale railway, struck near the Little Island Reef on or about 9 August 1872.[1] The vessel was in the charge of the pilot at the time.[2] She was refloated and brought in to Fremantle.

NOTES

1. V.G. Fall, *The Sea and the Forest* (University of Western Australia Press, Nedlands, 1972), p. 37.
2. Fremantle Harbour Master's Journal, 9 August 1872, Acc. No. 1056, Battye Library.

Hokitika

The iron barque *Hokitika* was built at Aberdeen, Scotland, in 1871 by Alex Hall and Co. On 7 October 1872, the new vessel left Newcastle, New South Wales, on her first trading voyage, laden with coal for Mauritius. After punching into westerly gales across the Bight, she arrived 20 kilometres off Cape Leeuwin at noon on 2 November. Captain Samuel Findlay set his course west-north-west into a south-westerly wind.[1] He had been round Cape Leeuwin ten times before, and probably felt confident of the area, despite the heavy swell that was running.

Although Robert Neill, the man at the helm, also had plenty of experience, he had been wrecked eleven times during the 20 years he had been at sea.[2] That afternoon, he made it an even dozen.

The barque was sailing at the rate of about 12 kilometres per hour when she struck an unseen reef at 3 p.m. Captain Findlay later said he thought that he was 25 kilometres off the land when the strike occurred. The mate sounded the pumps and found a metre of water. Captain Findlay stood in for the land immediately, but within 20 minutes the water had risen to 2 metres, and the vessel foundered between 45 and 60 minutes after first striking. The police

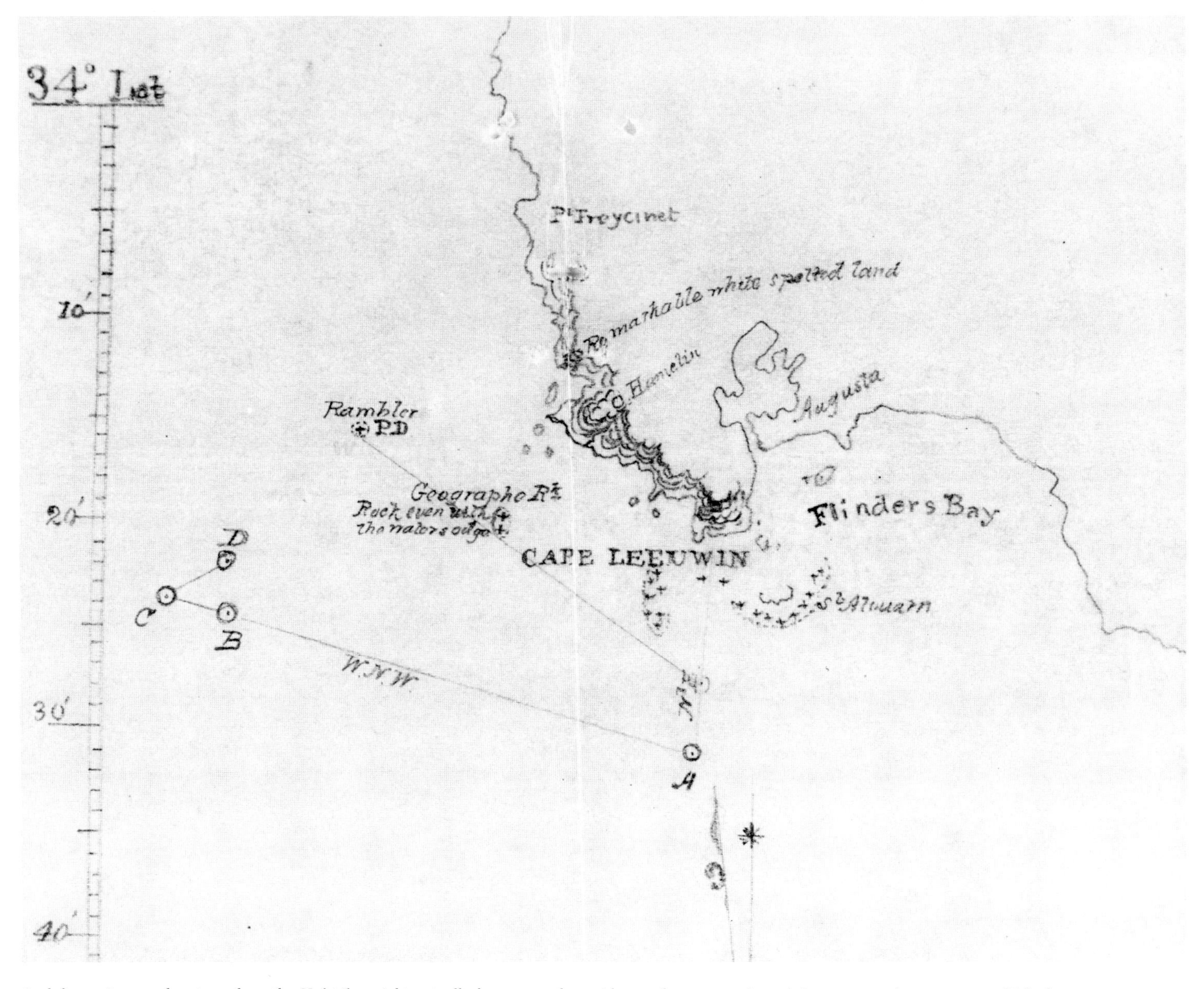

Archdeacon's map showing where the *Hokitika* might actually have gone down (the north-east route), and the west-north-west route which the crew claimed to have followed.

report stated that the vessel went down in 55 metres about 8 kilometres from shore.[3]

The eleven crew left for the land in the ship's two boats, and spent their first night ashore huddled by a small creek (probably Turner's Brook, about 13 kilometres north of Cape Leeuwin).[4] They started north in the boats at first light, and reached Busselton on the 5th.

A court of inquiry found that the *Hokitika* had struck an unmarked outlying rock many kilometres to the west of Cape Leeuwin, and that no blame could be attached to the master of the vessel. Today, we know that there are no rocks in the area Captain Findlay claimed that the vessel had struck. It seems clear that currents had carried the *Hokitika* far closer to the land than Findlay realized (or was willing to admit), and that the vessel struck the outlying Geographe Reef or Cumberland Rock, both of which are just 3 kilometres from shore. Allowing for 45 to 60 minutes of wallowing shoreward after she struck, it might be a reasonable guess to say that the *Hokitika* lies 2 or 3 kilometres north-east of Cumberland Rock, around 2 kilometres from shore. (See map on p. 20.)

Even at the time of the inquiry, Captain Findlay's stated estimate of his distance from shore did not convince everyone. Commander Archdeacon, in a private letter to the Colonial Secretary, argued that the conduct of that inquiry could only bring the Colony into disrepute, and he contended that, had the inquiry taken place before the Board of Trade, a very different finding would have resulted. He commented:

> At Fremantle I heard privately that the helmsman (who gave evidence) stated, after inquiry was over (upon being told I believe that he had lost the vessel) that the truth was that Captain and Mate were both drunk; and as this man had his clothes packed, anticipating an accident, one is apt to think there may be misinformation . . . Another report was also current at Fremantle, viz. that a herd boy, on the coast in vicinity of locality where the ship sank, stated that the vessel was so close in that he saw the people on her decks.[5]

The boats from the *Hokitika* were sold at Busselton. Nothing else was saved.

The 282-ton *Hokitika* (Official Number 64775), owned by merchants John and David Spence of Melbourne, had one deck, three masts and a round stern.[6] She was clincher built, with an iron frame, a female demi-figurehead and no galleries. Her dimensions were 40.4 metres by 7.6 metres by 3.7 metres.

In recent years, a mast section, made of wood, has been seen on a beach in the vicinity. If the wreck of the *Hokitika* is indeed closer inshore than Cumberland Rock, then it is likely that it will be found by skindivers in the not too distant future. During a Western Australian Museum expedition to the *Cumberland* wreck in March 1984, divers engaged in a swimline search between that wreck and the adjacent shoreline found coal on the seabed. This coal may have come from the *Cumberland*, or alternatively, from the *Hokitika*. When found, the *Hokitika* may prove to be in good condition and worthy of close attention by archaeologists.

NOTES

1. 'Preliminary Court of Inquiry', *Herald*, 23 November 1872.
2. *Ibid*.
3. Corporal R. Furlong's report, 6 November 1872, Police Records, Acc. No. 129, Battye Library
4. Captain Findlay, Report, November 1872, C.S.R. 727, fol. 153.
5. Commander Archdeacon to Col. Sec., 6 December 1872, C.S.R. 727, fol. 216.
6. Register of British Ships, Melbourne.

Minnie

The small pearling vessel *Minnie* (or *Minny*) was wrecked on Fortescue Island 11 kilometres out from the mouth of the Fortescue River, about 10 November 1872. The three men on board, George Forthcut, Liberty Joe and an Aborigine whose name was not given, reached the island with few provisions and remained there for about five weeks, hoping to see a passing vessel.[1] When they had nearly starved to death, the Aborigine swam to the mainland and walked to a pastoral station owned by a Mr Mackintosh.[2] (See map on p. 118.)

A boat was sent to the island, where Liberty Joe was found to be almost dead. However, the three men all recovered at the station. The Government Resident, Robert Sholl, rewarded the Aborigine by giving him a blanket, a shirt, a pair of trousers and a pound of tobacco.[3]

No details of the *Minnie* are available. It was probably the same vessel that had searched for the wreck of the *Nellie* earlier in 1872. It is possible that a schooner named *Minnie*, which is recorded as arriving at Fremantle from Shark Bay in 1879, is the same vessel.[4]

Another vessel of the same name, of 13 tons (Official Number 61033), was registered at Launceston in 1868.

NOTES

1. W.R. Piesse, Special Report, 23 December 1872, Police Records, Acc. No. 129, Battye Library.
2. W.R. Piesse, Report, 29 December 1872, Police Records, Acc. No. 129, Battye Library.
3. R.J. Sholl, Comment on Special Report, 22 March 1873, Police Records, Acc. No. 129, Battye Library.
4. *Inquirer*, 7 May 1879.

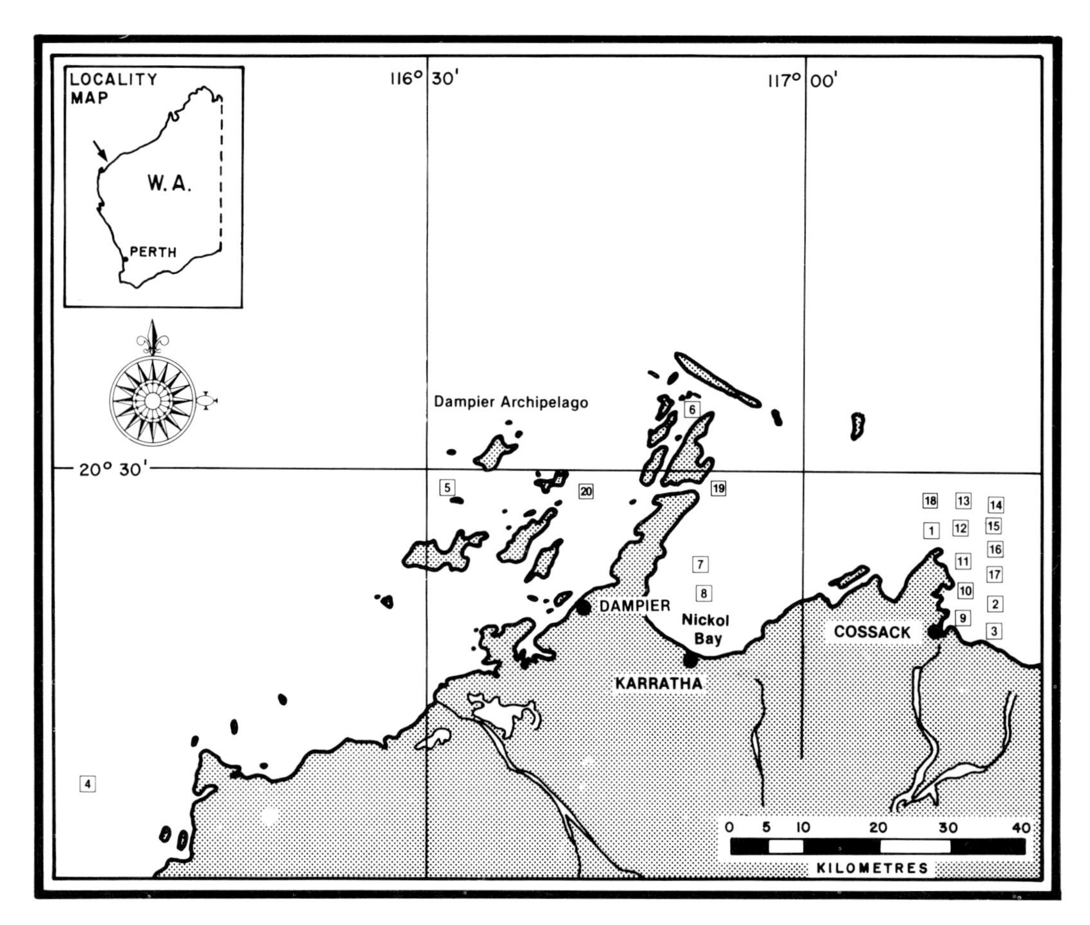

1 Unnamed Boat 1872
2 Unnamed Boat 1872
3 *Pilot*
4 *Minnie*
5 *Rosette*
6 *Speculator*
7 *Coquette*
8 *Nautilus*
9 *Mariano*
10 *New Perseverance*
11 Government Pinnace
12 *Sustenance*
13 *Champion*
14 *Bonnie Dundee*
15 *Conch*
16 *Square and Compass*
17 *Crest of the Wave*
18 *Flying Scud*
19 *Nellie*
20 *Industry*

Xantho

While the P. & O. Company was pioneering regular steam communiction between Europe and Australia with some of the largest vessels the world had ever seen, much smaller steamers were being placed on coastal runs within the Australian colonies by local entrepreneurs, with mixed degrees of success. Western Australia, the Cinderella colony, lagged slightly behind the others in most respects because of its isolation. Its first coastal steamer, the *Xantho* , was too old to be suitable for pioneering the coastal trade. It was introduced to the coast in 1872, and lasted less than a year before it was wrecked.

The schooner-rigged *Xantho* had been built as a paddle-steamer at Dumbarton, in the Scottish county of that name, in 1848, by William Denny and Company. The vessel was initially owned by a group consisting of a writer, a grocer, a draper and a merchant, who operated as the Anstruther and Leith Steam Shipping Company. The *Xantho* was certified by the Board of Trade to operate in rivers with partially smooth water.[1] In 1864, the certificate was changed to allow the vessel to be employed for excursions to sea, and the port of registration was changed to Wick.

During the period up to 1871, the *Xantho* had apparently been employed in the fashion of many other vessels of the later paddle-wheel era, on rivers and short coastal runs in sheltered waters. About that time, however, several interesting changes took place: the paddles were replaced by a propeller; the vessel was re-engined with a Penn steam engine originally intended for a gun-boat; and an entrepreneur bought the vessel to engage in the pearl shell fishery off the north-west coast of Western Australia.[2] It is surprising that such an old hull should have been re-engined, that the paddles were replaced by a screw (which would have increased stresses upon the old hull, built for sheltered waters) and that the vessel was, after this reduction in power, taken half way around the world to such an isolated and exposed coast.

The new owner of the *Xantho* was Charles Edward Broadhurst.[3] In 1864, Broadhurst had sailed with his family from Melbourne in the *Warrior*, as leader of the first party of shareholders in the Denison Plains Company, formed to pioneer grazing in the north of Western Australia. The party had intended to land at Roebuck Bay, but adverse winds caused them to land at

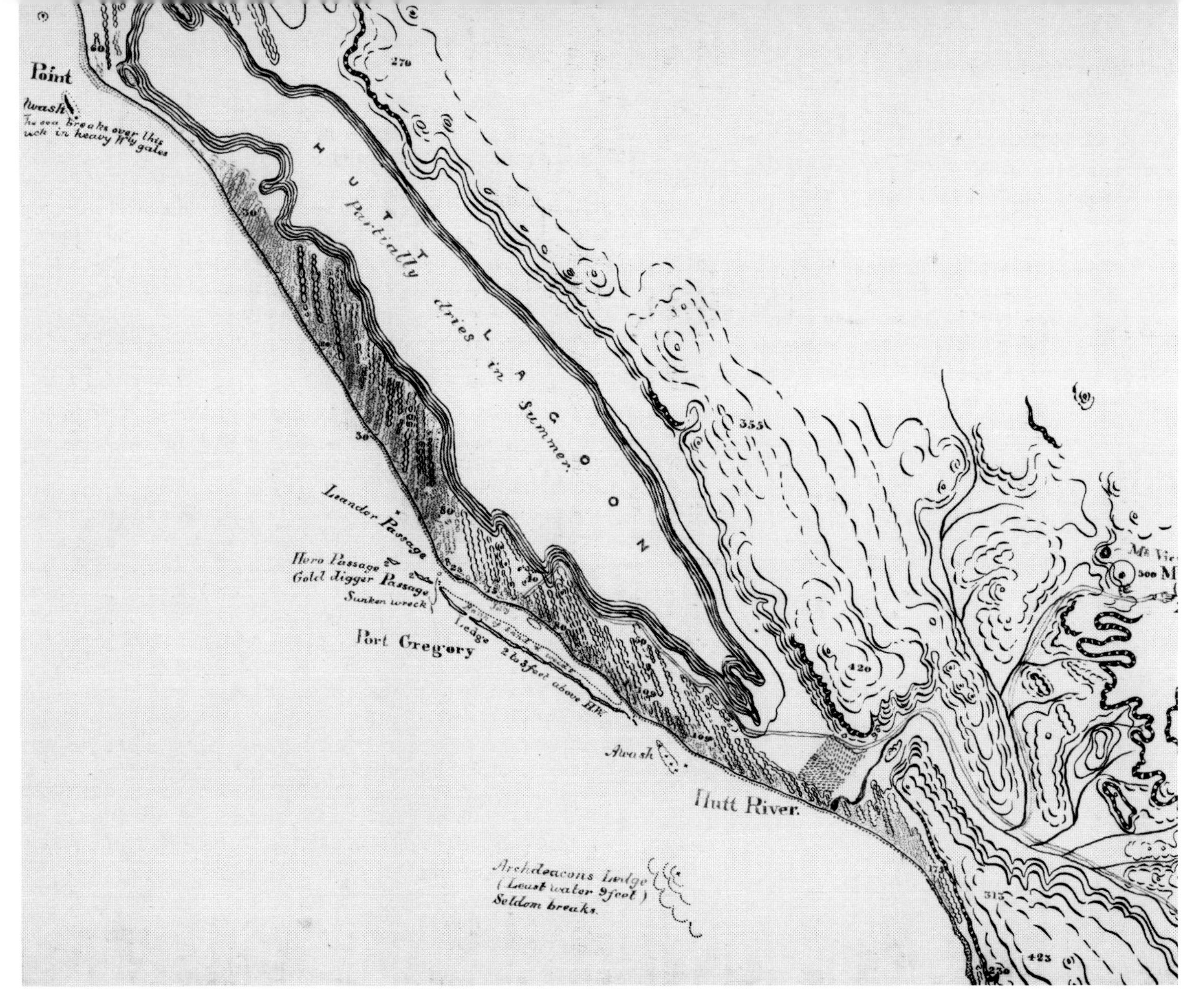

The wreck of the SS *Xantho*.

Nickol Bay, where Broadhurst obtained grazing leases. He was one of the first to systematically expand in the pearling industry, buying and fitting out a schooner for the purpose in 1870, obtaining Aboriginal prisoners for use as divers, and acquiring a Heinke diving apparatus in 1871.[4]

Broadhurst went to Glasgow to purchase the *Xantho*, bringing her out to Western Australia via the Suez Canal, Mauritius, Singapore and Batavia. After coaling at Batavia and engaging a number of divers, Broadhurst continued on to the pearl fishing grounds at Nickol Bay and finally reached Fremantle on 15 May 1872. On arrival at Fremantle, the *Xantho* had a crew of twenty, together with eight cabin and three steerage passengers, and a general cargo.

Although the vessel was generally welcomed at Fremantle, residents feared the possibility that it would divert the trade of the North West from its regular channels (through Fremantle) to Singapore, and thus deprive the southern centres of any benefits from the industries of the North West. The Government was urged to provide the *Xantho*'s owner with a pecuniary inducement to employ the vessel exclusively in the coasting trade.[5]

The *Xantho* left for Java in June with sixteen passengers and a cargo of spirits, tobacco, rams, potatoes, timber, shingles and sundries. It was expected to return in August to ply regularly between Champion Bay and Fremantle. At that time, colonists anticipated great improvements, because a mail contract between the southern settlements had just been awarded to the firm of Connor and McKay, and it appeared that there would soon be regular steam communication from the extreme northerly settlements to King George Sound on the south coast.

Events were soon to prove the *Xantho* incapable of performing the northern coastal run. After being detained for a long period at Champion Bay by tempestuous weather, she left on 2 July for Port Walcott and Batavia. Five of the Malay crew were reported lost between Roebourne and Banningarra.[6] The vessel was expected back at Geraldton some time in October, but was wrecked at Port Gregory on 16 November.

The *Xantho* had arrived at Port Gregory with a cargo of oil and wool, and had taken in 100 tonnes of lead ore intended for trans-shipment to the barque *Zephyr* at Fremantle. The vessel left Port Gregory, heavily laden, on 16 November, but about 10 kilometres out to sea, some of her fastenings started under the strain of the load, causing a heavy leak. The vessel was put about and headed back for the shelter of Port Gregory. Unfortunately, she did not quite make it, sinking in the very mouth of the harbour and becoming a complete wreck. Of the total cargo, a mere sixty-seven casks of oil were saved. The vessel was not insured. By the end of December, the engine room, cabin skylights and cabin companion-way had been washed overboard from the wreck. A diver reported that the main deck and bulwarks were also broken up and washed away. By February, the hull was full of sand and it was clear that the engine could not be salvaged. (See map on p. 53.)

Broadhurst's immediate problem was the payment of wages to the crew. He owed £500 in arrears and had agreed to return the Malays to Batavia at the completion of their service. The chief mate summonsed him for payment, but Broadhurst quietly left Geraldton the day before the case was to be heard. On the advice of the Resident Magistrate at Geraldton, Captain Denicke and most of the crew pursued Broadhurst to Fremantle, taking passage in the *Les Trois Amis*, a one-time steamer on the Swan River that had been converted to sail for the coasting trade.

The luckless crew arrived at Fremantle to find that Broadhurst had left that very day in the *Waterlily* for the North West, leaving no money and no instructions for the disposal of the wreck. Most of the crew found themselves utterly destitute.

Captain Denicke's remedy was to take it into his own hands to advertise the vessel for immediate sale, with the intention of appropriating the proceeds for himself and his fellow sufferers. The *Inquirer* listed the items for sale, and in so doing indicated what had been salvaged:

> Important Sale by Auction . . . The wreck of the above vessel, now lying in about 12 feet of water at Port Gregory, inside the reef, together with her furniture, tackle and rigging. Also the following articles stowed in a

Divers prepare the steam engine for raising.

> warehouse on shore: 1 complete set of sails, with running gear; 1 lower yard, topsail yard, fore gaff, and main boom and gaff, with gear complete. Also 1 bower anchor, 1 stream do, 81 fathoms chain, winch do, 2 boat's davits, 1 fist do, 2 lifebuoys, 2 boat's covers, manilla 3 inch line, coir warp, aneroid barometer, 2 thermometers, 3 salinometers, masthead and side lights, 4 cork fenders, rigging screws, 1 copper pump, 1 do hose branch, assorted blocks, gun, cartridges, blue lights, large ship's bell, portable forge, anvil, etc. Azimuth compass, tripod, 2 steering compasses, 1 patent log, engine room tools, 2 clocks, lamp, spy glass, complete set flags, new cooking apparatus, English flag and jack, 3 awnings, 1 thirteen feet dinghy etc. Also, at Fremantle: 1 first class ship chronometer by McGregor, Glasgow, and some maps . . .[7]

The auction went ahead amid threats of court action over title, and the sale only realized £180. As late as March 1873, Broadhurst still held hopes of raising the *Xantho*, but it would appear that little else was salvaged.

Charles Broadhurst later went on to become very successful in guano mining ventures, where he put to good use his earlier experience in employing indentured Asian labour.

In May 1979, a group of divers from the Maritime Archaeology Association of Western Australia went to Port Gregory to try to find the site of the *Xantho*. Armed with information the authors had extracted from the records of the court of inquiry, and assisted by local rock-lobster fishermen, they found the wreck after 3 days of searching. A large part of the hull has survived. The wreck lies at the Gold Digger Passage in latitude 28° 11.2′ south, longitude 114° 14.1′ east. The finders were given a reward of $3,000 by the Museum, and the Maritime Archaeology Association was contracted to survey the site. The Western Australian Museum has conducted several seasons of excavation on the stern section of the wreck, and the engine, raised in 1985, is currently undergoing restoration at the Museum's Conservation Laboratory.[8]

The iron clinch-built *Xantho* (Official Number 7802) had one deck, two masts, schooner rig, a round stern and a demi-figurehead of a woman. The dimensions of the vessel were 34.7 metres by 5.4 metres by 2.6 metres, and her original registered tonnage was 61.[9] The vessel was originally powered by a 60-horsepower steam engine.

NOTES

1. *Mercantile Navy List*, 1857.
2. *Inquirer*, 5 February 1873.
3. Henrietta Drake Brockman, Charles Edward Broadhurst, in *Australian Dictionary of Biography*, p. 233.
4. *Ibid*.
5. *Inquirer*, 29 May 1872.
6. *Inquirer*, 23 October 1872. Presumably, the men died of sickness or were washed overboard and drowned. The *Inquirer* does not elaborate.
7. *Inquirer*, 5 February 1873.
8. Mike McCarthy, 'SS *Xantho*: A 19th Century Lemon Turned Sweet', *Historical Archaeology*, Special Publication No. 4, 1985, pp. 54–6. See also Mike McCarthy, 'Treasure from the Scrapheap', *Australian Sea Heritage*, Spring, pp. 22–5.
9. Board of Trade Transcripts 108/67, Public Record Office, England.

The engine manufacturer's plate after cleaning.

Fanny Nicholson

The barque *Fanny Nicholson* was built at Hartlepool, in the English County of Durham, and brought to Australia in the 1860s. During the early 1870s, she operated as a whaler, based in Hobart and fishing along the south coast of Australia. She visited Albany several times in this capacity.[1]

In April 1872, the *Fanny Nicholson* set out on a whaling voyage from Hobart, under the command of Captain Gaffin, and arrived at Albany with a cargo of oil on 21 November. At midnight of the 22nd during a heavy gale from east-south-east, the vessel parted from both her anchors and went ashore in Frenchman's Bay. Some of the crew arrived at the town at 2.30 a.m.; the remainder of the twenty-six men, finding the seas too rough to cross to Albany, went in two whale boats to Rabbit Island, where they were rescued later in the day.[2] By 9 December, it was reported that the *Fanny Nicholson* had become a total wreck. The barque *Free Trader* salvaged oil, rigging and whaling gear from the wreck early in 1873.

The position of the wreck was plotted by Commander W. E. Archdeacon during his survey of the area in 1877 in the ship *Runnymede*.[3] The wreck can still be seen, partly buried in sand on the seabed. (See map on p. 126.)

The 285-ton *Fanny Nicholson* (Official Number 23706) had a wooden frame, one deck with a poop, three masts, a standing bowsprit, an elliptical stern and a figurehead of a woman. Her dimensions were 36 metres by 7.7 metres by 4.6 metres.[4] In 1872, the vessel was owned by William Andrews, of Sydney.[5]

NOTES

1. See Report for the Port of Albany, 15 December 1871, C.S.R. 702, fol. 150, and Report for the Port of Albany, 21 November 1872, C.S.R. 728, fol. 108.
2. Report of Corporal J. Campbell, 26 November 1872, Police Reports, Acc. No.129, Battye Library.
3. Archdeacon, Princess Royal Harbour Chart No. 27. See Bruce Melrose, 'Wreck of the Month', *Underwater Explorers Club News* (May 1967), p. 12.
4. Register of British Ships, Sydney.
5. *Mercantile Navy List*, 1872.

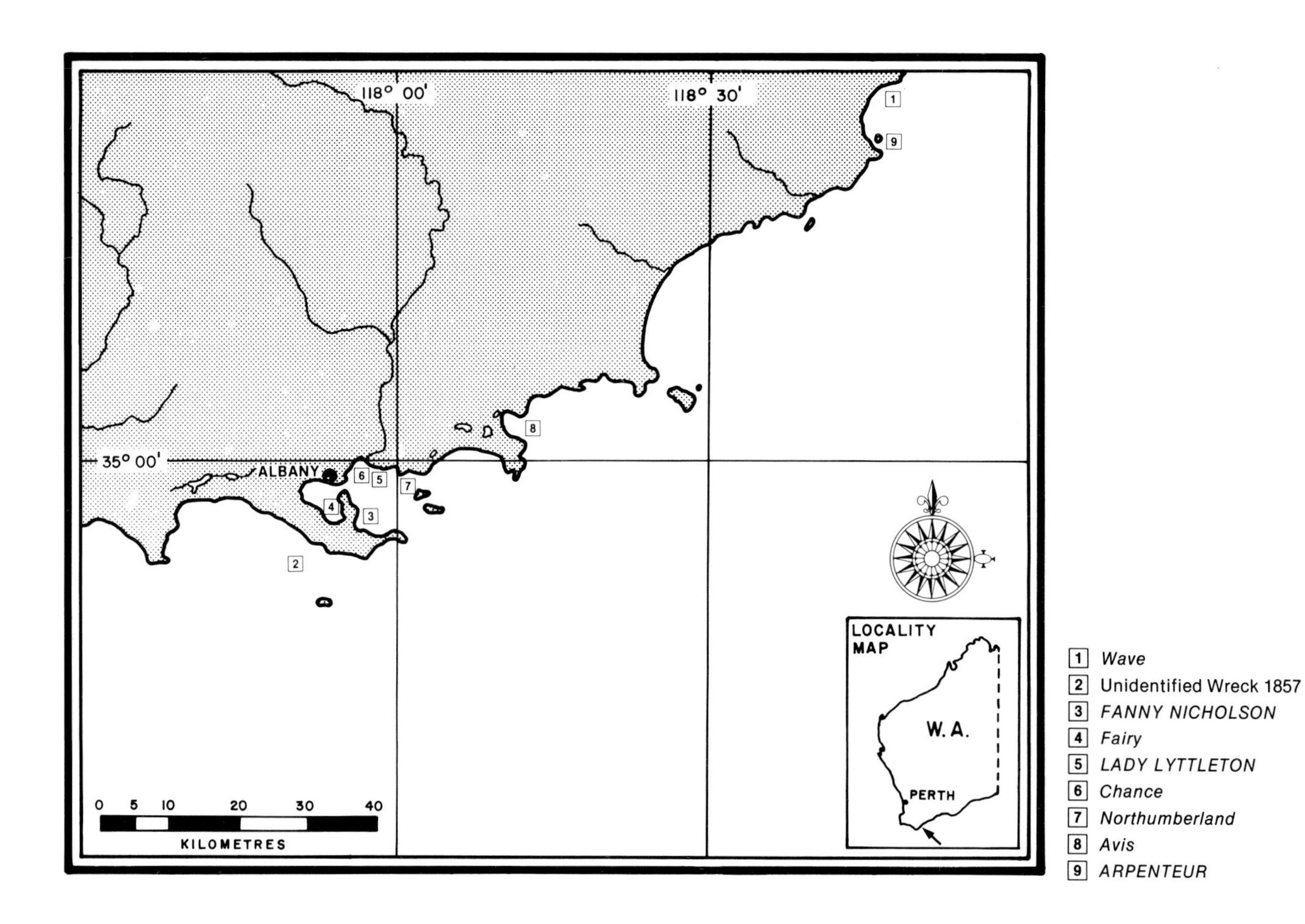
118° 00'
118° 30'
35° 00'
ALBANY
1
2
3
4
5
6
7
8
9
0 5 10 20 30 40
KILOMETRES
LOCALITY MAP
W. A.
PERTH
1 Wave
2 Unidentified Wreck 1857
3 FANNY NICHOLSON
4 Fairy
5 LADY LYTTLETON
6 Chance
7 Northumberland
8 Avis
9 ARPENTEUR

Twinkling Star

The Fremantle firm of Batemans bought the schooner *Twinkling Star* in March 1868 for £1,000.[1] The vessel had been owned by shipbuilder Benedict Von Bibra in Calcutta and made several voyages between there and Australian ports. But Batemans were more interested in the coasting trade.

During the following years, Batemans used the *Twinkling Star* to supply the Champion Bay settlers with building materials, beverages and manufactured goods, in exchange for farming produce such as wheat, barley, bran, hides, flour and oats. Passengers were also carried between the two ports of Geraldton and Fremantle. This service was interrupted by occasional longer voyages, such as one to Batavia in 1869 and another to Melbourne in 1871.

The schooner was wrecked on a voyage from Champion Bay to Fremantle in January 1873. She was commanded by George Long, assisted by five crew, and carried eight passengers, as well as a cargo of lead ore, hay and flour. After thirteen days of beating against southerly winds, the *Twinkling Star* had passed the latitude of Fremantle. At noon on the 30th, Captain Long placed the vessel 110 kilometres off the land, but he had the longitude only by guess.[2]

That afternoon, he steered east-half-north (84.25°) at the rate of about 8 kilometres per hour, and hove the lead at 11 o'clock. However, 15 minutes before midnight, the man on the lookout saw land on the lee, and the *Twinkling Star* struck a reef on the west side of Garden Island. A passenger later stated that the reef was about 1.2 kilometres offshore, opposite an elevated part of the island (probably Mt Haycock).[3]

Efforts to work the vessel off the reef were to no avail, so the passengers were landed on Garden Island. Two days later John Bateman took a cargo boat to render assistance, but found that the seas breaking over the *Twinkling Star*'s decks had carried away the roundhouse and were threatening to break the hull apart. (See map on p. 242.)

Nevertheless, he managed to salvage the masts, spars, rigging, and part of the cargo. Samuel Solomon reported receiving into his store a boatload of forty-two bags of lead ore from the wreck, so presumably the vessel lay in an accessible position on the reef platform adjacent to the shore.[4] The *Twinkling Star* was not insured.

At the court of inquiry, Captain Long admitted that he had only guessed his longitude during the voyage. He argued that it was usual on the west coast for masters of vessels to calculate longitude by 'rule of thumb'.[5] The court censured Long for standing inshore after dark, for not heaving the lead more frequently, for trusting to obtaining a sighting of the Rottnest light, and for not ascertaining the longitude by appropriate means. He should have checked his longitude either by chronometer or by dead reckoning.

In modern, times the wreck site has eluded searching skindivers. Although a small vessel, the survivors' descriptions closely define the locality of the wreck, and the lead ore cargo could be expected to remain conspicuous.

The *Twinkling Star* (Official Number 49320) was built at Subpore Howrah, Calcutta, in 1866. She was a 59-ton two-masted schooner with a round stern, and her dimensions were 19.3 metres by 4.9 metres by 2.2 metres.[6] The vessel had been re-coppered in 1872.

NOTES

1. *Inquirer*, 4 March 1868.
2. George Long, evidence given at the Inquiry into the loss of the *Twinkling Star*, held on 14 February 1873, C.S.R. 735, fol. 102.
3. W. Simmons, evidence given at the Inquiry, C.S.R. 735, fol. 105.
4. S. Solomon to J. Absolon, 4 February 1873, Habgood Papers, 813A, Battye Library.
5. A 'rule of thumb' is a procedure derived entirely from practice, without any basis in scientific knowledge.
6. Register of British Ships, Fremantle.

Sea Spray

In February 1873, the Colony lost the largest of its pearling vessels, the 31-ton schooner *Sea Spray*. The vessel had left Flying Foam Passage for Condon, leaking badly, and on entering Condon Creek, had struck upon a bank. The crew could not get the pump to work and had to bail the schooner out with a bucket.[1]

When she left Condon the *Sea Spray* carried twenty-eight people, consisting of five Europeans, three Aborigines and twenty Malays. On deck were forty sheep and two dinghies. The vessel was last seen afloat off Port Hedland, beating to the westward against a strong head wind and 'flying light' because, with the exception of her living cargo, she had on board only fifteen hogsheads of water.[2]

It seems probable that the *Sea Spray* foundered. On 16 March, Captain Black, of the *Black Hawk*, found the *Sea Spray*'s winch. One of his crew saw the wreck itself. Charles Hanson later stated:

> On the 15th March I left the schooner *Black Hawk* in Port Hedland with a boat and crew. I went to the east entrance of Port Hedland and I found two water casks, one of them marked S.S . . . I then proceeded in one of the ship's boats in search of the wreck of the *Sea Spray*. I went to the N.W. passage of Port Hedland and the following morning I sighted the wreck of a two masted vessel, about 7 miles off shore about W.N.W. of the entrance. She appeared to be aground. I believe one of the masts had given way by being unstepped. It had fallen against the other mast which was standing which, I believe to be the main mast. I believe there was only a part of the vessel there. I believe the bow of the vessel was gone, that it had broken away. I proceeded that afternoon to Sandy Island and picked up one hogshead there marked S.S. I thought I would see the wreck better from the Island but when I got there the flood tide had washed the wreck away . . . . From where I first saw the wreck I think it would be about N.E. from Sandy Island.[3]

Other bits of flotsam were found, but the wreck was not seen again. The twenty-eight men on board must have all drowned. (See map on p. 32.)

The *Sea Spray* (Official Number 36546) was built at Fremantle in 1862, and was owned by John Shea. It was carvel built, with a wood frame, and measured 15.2 metres by 4.6 metres by 2.1 metres.[4] The vessel was constructed with one deck, two masts and a square stern.

NOTES

1. *Herald*, 10 May 1873.
2. *Ibid*.
3. Charles Hanson, statement made at Flying Foam Passage on 4 April 1873, C.S.R. 752, fol. 79.
4. Register of British Ships, Fremantle.

Premier

The schooner *Premier* was built at Perth in 1869 for boatbuilder Charles Watson and pastoralist James Brockman. The vessel was employed in the coasting trade on the west coast. Early in 1873, Brockman had the *Premier* lengthened from 22.8 metres to 26.2 metres, probably in response to the increasing volume of coastal trade.[1]

The *Premier* sailed from Fremantle on 16 June 1873 and arrived at Port Irwin 12 days later. On 1 July, while anchored at Port Irwin, the vessel began to drag two anchors before a strong north-westerly wind. A third anchor was let go, but as the wind strengthened to a moderate gale, the schooner commenced dragging again, and went ashore on rocks at 4 p.m.[2] She lay there in about 2 metres of water while the false keel was ground away and the rudder smashed from its hinges. (See map on p. 44.)

Two weeks later, the cargo of telegraph posts had been recovered, but the *Premier*'s bottom had been broken. An observer lamented that, with proper gear and good tackle, the vessel could be hauled ashore and repaired, but that no such equipment was available. An inquiry into the loss of the vessel concluded that poor holding ground (the seabed consisted of flat rock) was responsible for the *Premier* being wrecked.[3]

The *Premier* (Official Number 61092) was built of jarrah as a 51-ton two-masted schooner, with one deck, a square stern and the dimensions 22.9 metres by 4.6 metres by 1.9 metres.[4] During the refit early in 1873, she was given a round stern and had her dimensions increased to 26.2 metres by 5.5 metres by 2.4 metres.

NOTES

1. Register of British Ships, Fremantle.
2. Henry McRanls, evidence at Inquiry into the stranding of the schooner *Premier* at Port Irwin, C.S.R. 735, fol. 133.
3. Inquiry, C.S.R. 735, fol. 133.
4. Register of British Ships, Fremantle. See also Board of Trade Wreck Register, 1873, National Maritime Museum, England.

Wild Wave

The brig *Wild Wave* (formerly the *China*) was purchased in 1872 by the stock-owner and entrepreneur George Howlett, of Roebourne, who had the vessel's registry transferred to Melbourne and employed her in trade with Western Australian ports.

Howlett enthused about his acquisition and plans in a letter to his Fremantle agent, John Absolon:

> I have just purchased a fine vessel called the *Wild Wave*. She carries about 180 tons and is about the fastest and best vessel of her size that comes to Melbourne. I have also got a partner to join me. He finds £3,500 pounds in cash and stores, vis £1,000 cash the rest in stores, he has also another vessel of about 69 tons called the *Victoria* which he brings into the concern, so that we shall have in all three vessels to carry on the trade of the place. My partner's name is Crouch and he is to place £500 to the credit of the firm at Perth . . . I shall be leaving here for Port Walcott about the 28th inst and I hope I shall be in time to bring all the wool to Fremantle in one trip. I shall have on board about a dozen thoroughbred mares and horses for Nickol Bay, and also some rams and nine or ten passengers and quite a full cargo all our own.[1]

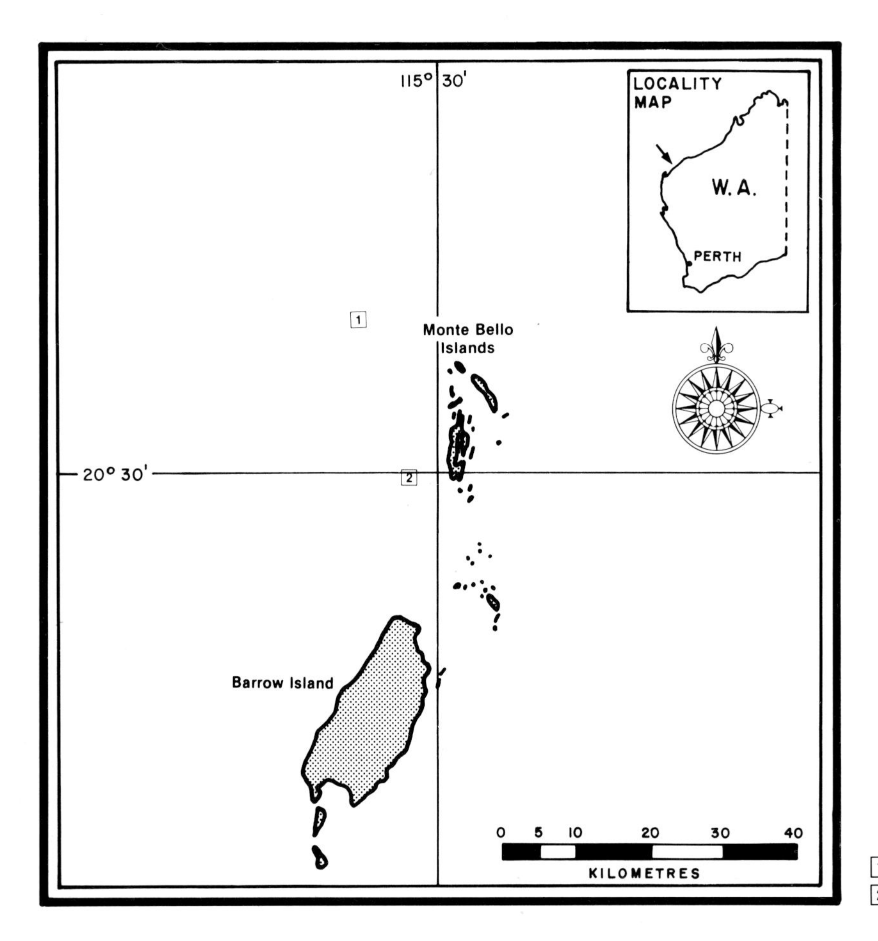
115° 30'
LOCALITY MAP
W. A.
PERTH
Monte Bello Islands
20° 30'
Barrow Island
0 5 10 20 30 40
KILOMETRES
1 TRIAL
2 Wild Wave

On her first visit to the west coast, in 1872, the *Wild Wave* sailed to Fremantle from Port Darwin via Port Walcott and Champion Bay. Calms detained the vessel on the voyage past the Kimberley coast. Howlett was on board, and to occupy his time, he followed the immense flights of sea birds, which seemed to be converging on a particular point. The birds led him to Browse Island, where he took guano samples.

Howlett sent the samples to Melbourne on the *Wild Wave* for testing, and they proved to be extremely rich in phosphates. With thoughts of exploiting the deposits, he sent the *Wild Wave* back to Fremantle in July 1873, and arranged for a voyage to Singapore via Port Walcott, under the command of Captain Edward Fothergill.[2]

The *Wild Wave* sailed from Fremantle on 25 August with twenty-seven passengers, including Fothergill's wife and two children, Howlett, and a crew of fifteen Malays. At noon on the 30th, the ship was in latitude 25° 55′ south, with Point Cloates bearing north-east 50 kilometres away. The vessel sped past North West Cape and Barrow Island. Captain Fothergill thought the brig would be well to the west of Barrow Island, but a current had brought it quite close to land. The studding sails were taken in and a lookout posted on the foreyard at midnight. However, the brig was going at 13 kilometres per hour 1½ hours later, when the officer of the watch saw breakers ahead.[3]

The brig struck and at once filled with water. The boats were made ready to leave the ship, and at daylight the unhappy crew saw the Monte Bello Islands 13 kilometres to the east. They made for the southern end of the Group and set up camp, probably on Hermite Island. Lockier Burges in his reminiscences states that the *Wild Wave* was lost in the vicinity of 'Big Sandy Island', but that name is not listed in the *Gazetteer*.[4]

On 6 September, Captain Fothergill set out with his family and one of the passengers in a boat to seek assistance. He fell in with the schooner *Mary Ann* at Flying Foam Passage and dispatched that vessel back to the Monte Bellos, while he pushed on to Roebourne. The wreck was later sold at auction for £200 to Cossack importer Charles Crouch, who salvaged most of the cargo of sandalwood and mining equipment.[5]

Nine of the Malay crew were engaged by a pearler named McKay to work on one of his vessels in return for the ordinary wages and a passage back to Surabaya at the completion of the season. They were paid for the first month, but then taken to Shark Bay, where they were ill-treated and given no food. At the end of the second month, they ran off and lived in the bush, nearly starving, for about 2 months, and were then engaged by the pearler Charles Broadhurst.[6] Broadhurst later took money from them and refused to return it.

The loss of the *Wild Wave* prevented Howlett from entering into the mining venture alone, so he formed a company. The company surveyed the deposits, which they roughly estimated at 200,000 tonnes, covering the island at a depth of between 1.0 and 1.8 metres.[7] A lease was taken out with the Western Australian Government, and exploitation of the deposits began in 1876.

The *Wild Wave* (Official Number 43302) was built at Abenraa in Denmark in 1858 by Peter Lund. The 180-ton vessel, measuring 31.4 metres by 7.4 metres by 3.9 metres, had one deck with a break, two masts, a round stern, a wooden frame, and a snake's head figure.[8]

Some confusion has been created by the Board of Trade Wreck Register, which wrongly lists two brigs of the same name wrecked on the Monte Bellos in 1873. The second vessel (Official Number 32299), 191 tons and registered in Sydney,[9] was still included in the *Mercantile Navy List* of 1882.[10]

NOTES

1. George Howlett to John Absolon, 1 May 1872, Habgood Papers, 813A, Battye Library.
2. Register of Arrivals at Fremantle.
3. Captain Edward Fothergill, evidence at the Inquiry held at Cossack, 18 October 1873, C.S.R. 736, fol. 128.
4. L. C. Burges, *The Pioneers of the Nor'-West, Australia* (Constantine and Gardner, Geraldton, 1913), p. 32
5. *Inquirer*, 29 October 1873.

6. Lieutenant Thomas Suckling (Commander H.M.S. *Renard*) to Governor William Robinson, Fremantle, 16 April 1875, in Correspondence relative to state of affairs on the North West Coast and the treatment of the Malay and other labourers employed in the pearl fisheries, *Votes and Proceedings* (Perth, 1875).
7. *Inquirer*, 5 July 1876.
8. Register of British Ships, Melbourne.
9. Board of Trade Wreck Register, 1873, National Maritime Museum, England.
10. *Mercantile Navy List*, 1882.

Robert Morrison, Little Eastern, May, Annie, Two Sons, Mazeppa, Dawn, Macquarie, Rose, Vixen, Alma and Anna

The casualties to shipping at Fremantle in 1873 surpassed anything hitherto experienced at the port. A cyclone over the weekend of 6–7 September left a fleet of vessels lying damaged on the beaches. (See map on p. 242.)

The weather became boisterous on the Saturday, and at 8.40 p.m., the 533-ton ship *Robert Morrison*, a regular wool trader, showed blue distress lights and rockets, and fired her guns in quick succession. The *Robert Morrison* was anchored in Gage Roads near Minden Rock, close to Fremantle township. Captain Thomas Coates would have preferred to have moored her in the winter anchorage at Garden Island, but the vessel's draft (she drew 5.3 metres) was too deep for the channel in.[1]

In the circumstances, Captain Coates would have been wise to have got the vessel under weigh and headed out to sea when the barometer glass fell on the Saturday. But he and some of the crew had spent the day on shore, and by the time he returned to his ship at 11 p.m., it was bumping on the bottom, with 1.2 metres of water in the hold. He set the reduced crew to work at the pumps, but at 8.30 a.m. on the Sunday, they gave up the task. Captain Coates had heard a crack, which he imagined to be the vessel's back breaking, so he took all the men ashore.[2]

An enterprising Fremantle boatman, William Back, heard that the *Robert Morrison* had been abandoned and set off in a whale boat with eight seamen to take possession of her. Back's men were busy at the pumps when Captain Coates returned at 10 a.m. to dispute possession. Soon afterwards, the vessel drove towards the beach and grounded a short distance off the old jetty.[3] Back returned the vessel to Captain Coates the following day for £500!

Meanwhile, the storm had been causing havoc among the other anchored vessels. At 9.20 p.m. on the Saturday, the 12-ton boat *Little Eastern* parted her moorings and went ashore near the watering jetty. At 1.10 a.m. on the Sunday, the cutter *May* drove ashore in the same place, carrying away her rudder and breaking her bulwarks in the process. At about 6 a.m., the cutter *Annie* drove ashore near the watering jetty, a plank being stove in and some of her bulwarks carried away. At 7.55 a.m., the cutter *Two Sons* drove ashore opposite the Water Police Station, without any damage. At the same time, the cutter *Mazeppa* parted her cable and was beached. Six passengers and four Aboriginal prisoners scrambled through the surf to safety. At 11 a.m., the schooner *Dawn* was beached near the watering jetty, demolishing part of the jetty with her stern. The schooners *Macquarie* and *Rose* (94 tons) drove ashore close by, only a trifle damaged. The *Vixen* and the *Alma*, two small boats intended for the Shark Bay pearl fishery, drove ashore and were quickly broken up. At the Rockingham Timber Station, the 144-ton barque *Anna* parted her cable and drove ashore.[4]

However, the situation was not as bad as it seemed. Two days later, the *Macquarie* had been floated, with only slight damage to the rudder. The *Two Sons* and the *Little Eastern* were refloated without much trouble. The *Anna* was soon refloated and, when surveyed, was found to have no apparent damage beyond straining (the *Anna* left for New Zealand with a cargo of jarrah and was wrecked *en route* at King Island on 26 October).[5] The *Rose* had to discharge its cargo of timber and lost its rudder and false keel. It was not insured, so the owner suffered substantial repair costs. The *Dawn*'s master also had to discharge his cargo of sandalwood, but the only damage was a lost rudder. The *Annie* remained on shore for some time, but as it was a tiny 7-ton cutter, the costs were presumably not great. The *Mazeppa*'s lead ore and hay were discharged without much damage, and the vessel was repaired for only £25.

The *May* was substantially damaged. The vessel's master, John Vincent, estimated the cost of damage to the cargo at £300, and the damage to the vessel at £130.[6]

Nothing further was mentioned about the two wrecked pearling vessels *Vixen* and *Alma*. They were small and probably did not constitute a large financial loss. Nevertheless, it would have been interesting if more details had been given about the construction of these craft. Large new beds of pearl oysters were being opened up at Shark Bay at that time, and the approach to the industry there (the shell was dredged) was quite different to the approach taken to the larger shell further north, where divers collected the shell.[7]

The *Robert Morrison* was eventually refloated and taken across to Garden Island in November 1873. Four months later, it was decided to call tenders for repairing the vessel with Bateman and Pearse undertaking the job for £1,600. In April 1874, it was announced that the *Robert Morrison* would be sailing for Singapore with a batch of horses.[8] It was by then outside the wool season, so she could not carry her normal cargo to London.

NOTES

1. Niel McEachern (Port Pilot), evidence given at Preliminary Inquiry into the stranding of the *Robert Morrison*, held on 9 September 1873, C.S.R. Customs 736, fol. 143.
2. Thomas Coates, evidence at Preliminary Inquiry into *Robert Morrison*, C.S.R. 736, fol. 151.
3. *Inquirer*, 10 September 1873.
4. *Ibid*.
5. Board of Trade Wreck Register, 1873, National Maritime Museum, England.
6. John Vincent, evidence given at Inquiry on 22 September 1873 into the stranding of the *May*, C.S.R. 735, fol. 153.
7. *Inquirer*, 10 September 1873.
8. *Inquirer*, 24 April 1874.

Amy

The town of Dongara was visited with an unexpected storm on 10 September 1873. The schooner *Amy* had been loading with sandalwood for Champion Bay when the nor'-wester struck. She dragged her anchors and collided with great force with the jetty. The vessel was dashed towards the shore and lodged on the beach, with her stern several feet under the jetty.[1]

A week later, it was reported that the *Amy* was being repaired and was expected shortly to be refloated.[2] The Register of British Ships indicates that the vessel continued operating until at least 1890.

The 32-ton schooner *Amy* (Official Number 61102) was built at Perth in 1869.[3] The wood-framed vessel had two masts, one deck, a square stern and the dimensions 17.1 metres by 4.3 metres by 2.1 metres.

NOTES

1. *Inquirer*, 17 September 1873.
2. *Inquirer*, 22 September 1873.
3. Register of British Ships, Fremantle.

Emilienne, Sea Ripple and Annie Beaton

Another three vessels were stranded at Fremantle on 28 October 1873. During a severe gale, the schooners *Sea Ripple* and *Annie Beaton* drifted ashore on the south beach, and the French barque *Emilienne* was driven on the rocks under Arthur Head at the mouth of the Swan River.[1] Neither the *Sea Ripple* nor the *Annie Beaton* suffered any damage, but the *Emilienne* was at the time considered wrecked.

The *Emilienne* lay on the rocks with 2.4 metres of water in the hold until the Governor came to the rescue by supplying a fire-engine pump and the labour of fifty prisoners.[2] The vessel was floated to the new jetty, where she again filled with water and posed an obstruction to other shipping. In March 1874, the vessel was refloated and towed across to Garden Island by two of George

Randell's steamers. There, she was hove down and repaired, finally leaving port in November 1874, after a stay of over 12 months.[3]

NOTES

1. *Inquirer*, 5 November 1873.
2. Captain Corseul to Governor Frederick Weld, Fremantle, 5 November 1873, C.S.R. 736, fol. 56.
3. *Herald*, 4 November 1874.

Bateman's well-known trader the *Sea Ripple*.

Annie Beaton

The schooner *Annie Beaton*, trading between Fremantle and the North West settlements, grounded on one of her anchors at Cossack early in 1874. The hull was holed, but the leak was repaired, and in April 1874 it was reported that she had arrived in Roebourne with a valuable cargo of shells.[1]

NOTE

1. *Herald*, 16 May 1874.

Rover

The *Inquirer* records that Charles Broadhurst's cutter *Rover* went down in the South Passage at Shark Bay in March 1874, and it was thought that the vessel would become a total wreck.[1] However, a cutter of that name was listed as pearling east of Tien Tsin harbour in December 1875, so it may be that Broadhurst's vessel was refloated. (See map on p. 250.)

A 14-ton jarrah schooner named *Rover* (Official Number 61109) was built at Port Walcott in 1872. It was two-masted, with one deck and a round stern, and measured 12.5 metres by 4.1 metres by 1.5 metres. The Register notes the pastoralist-pearlers John Edgar and Alexander Richardson as the owners.[2] Although remaining on the Register until 1908, it was not traced after 1873.

NOTES

1. *Inquirer*, 13 May 1874.
2. Register of British Ships, Fremantle.

Chalmers

The British ship *Chalmers*, carrying sugar from Mauritius to Fremantle, struck the Murray Reef on 19 March 1874 and was wrecked. A preliminary investigation into the causes of the loss of the vessel resulted in formal charges being preferred against the master, Captain William Alexander.[1] (See map on p. 242.)

The *Chalmers* had left Mauritius on 7 February, and all went well until about 11 p.m. on 19 March, when the vessel struck. Captain Alexander said that he had seen a bush-fire on the mainland and steered for it, believing it to be the Rottnest light. He was charged as follows:

> 1st That you caused the wreck of the *Chalmers* through incompetency and negligence.
>
> 2nd That you caused the vessel to stand on a course E by S after the light was sighted at 8 p.m. on 19th till she struck between 11 and 12 at night without causing the lead to be hove.
>
> 3rd That after the vessel grazed the bottom between 11 and 12 of the 19th instead of pulling the helm hard down at once you continued on the same course.
>
> 4th That after the vessel struck the second time and got clear with water of 2½ or 3 fathoms, in fine weather without any sea, you neglected to bring to, or anchor the said vessel.
>
> 5th That after the vessel became fast on a sand bottom, and was making water at the rate of 1 inch per hour, you neglected to pump the vessel.
>
> 6th That after the vessel was hove afloat by the kedge you neglected to anchor the vessel or sound with a boat to ascertain if a clear passage out was practicable, before cutting the warp.[2]

The court found that 'the whole of the conduct of the master once striking the bottom is utterly inexcusable on any supposition except that he lost his presence of mind and judgement entirely'.[3] Captain Alexander's certificate was cancelled. He appealed, without success, the court deciding:

> . . . Viewing the whole case we cannot pass over the fact that every action taken by the Master from the sighting of the light almost appears to have been premeditated. It is our duty however to state that no evidence was

> adduced that would warrant us in assuming for a moment, that there was any intention to lose the vessel.[4]

Meanwhile, the hull of the wrecked *Chalmers* was sold at auction to Messrs J. & W. Bateman for £19. The cargo was destroyed, but much of the ship's equipment was saved, Batemans employing three lighters to dismantle the wreck.

The wreck was located in May 1975 by skindivers Graham Anderton, Jeremy Prince and L. Gillett, all members of the Living Water Skindiving Club. The finders were given a reward of $75 by the Trustees of the Western Australian Museum. The wreck lies in 4–7 metres of water on the inshore side of a small breaking reef at the southern end of the Sisters Reef, in latitude 32° 22.0′ south, longitude 115° 41.2′ east. The hull timbers are covered by a mound of ballast stones.

The *Chalmers* (Official Number 12542) was a ship-rigged vessel of 594 tons, built at Sunderland in 1851 by and for James Laing. Her dimensions were 40.1 metres by 9.1 metres by 6.2 metres. Although 23 years old at the time of her loss, the vessel had retained its A1 classification with Lloyds.[5] The ship was sturdily built, with English oak stem and stern-post, and nine pairs of iron hanging knees. The exterior, including the flat of the upper deck, was fastened with yellow metal to the entire exclusion of iron.[6]

NOTES

1. *Herald*, 28 March 1874.
2. Report on William Alexander's Petition by the re-assembled Court, Fremantle Customs House, 12 April 1875, C.S.R. 813, fol. 89.
3. L. W. Clifton to Col. Sec., Fremantle, 7 April 1874, C.S.R. 813, fol. 72.
4. Report on Petition, C.S.R. 813, fol. 94.
5. *Lloyds Shipping Register*, 1873–4.
6. Lloyds Survey Register, No. 4531, National Maritime Museum, England.

Amelia

On 20 March 1874, the cutter *Amelia* capsized at Shark Bay, almost drowning Charles Broadhurst and another man. A fortnight later, Broadhurst was still suffering from his immersion.[1] The same vessel had capsized on a previous occasion, drowning two pearlers. (See map on p. 250.)

The *Amelia* appears not to have been registered at Fremantle, and it is not known whether the vessel was subsequently refloated.

NOTE

1. *Inquirer*, 13 May 1874.

Sea Bird

The schooner *Sea Bird* was built at the Vasse in 1865 for Henry Yelverton, who employed her for carrying timber. He sold the vessel to Wallace Bickley and William Richardson of Bunbury, and they in turn sold to George Shenton, of Perth. When Shenton died in 1867 (one of the victims of the *Lass of Geraldton* sinking), the schooner became the property of Charles Crowther, of Geraldton.

The *Sea Bird* left Fremantle for the north on 11 June 1874 with a general cargo and ten passengers, the majority of whom were ticket-of-leave men. The master, Charles Hanham, had been given command of the vessel at the last moment, and he imprudently left port without a log line for establishing his speed.[1] Fine weather lasted until midday on the 13th, when Hanham saw what he thought to be Mount Lesueur (close to Geraldton).

From that time onwards, the glass fell and the wind rose. The *Sea Bird* beat about for 2 days, trying to keep off shore, but for a time the wind was so strong that all sail other than the jib had to be taken in. Hanham, in desperation, ran the vessel in behind a reef 5 kilometres from shore and anchored for the night. He had no idea of his latitude. The *Sea Bird* was an unusually flat-bottomed vessel and had made more lee-way than Captain Hanham realized.

On the 16th, the wind shifted to north-west and the *Sea Bird* began to drag her anchors. As the vessel drew perilously near to a reef, Hanham slipped the mooring chains overboard, and for the safety of his passengers, ran the *Sea*

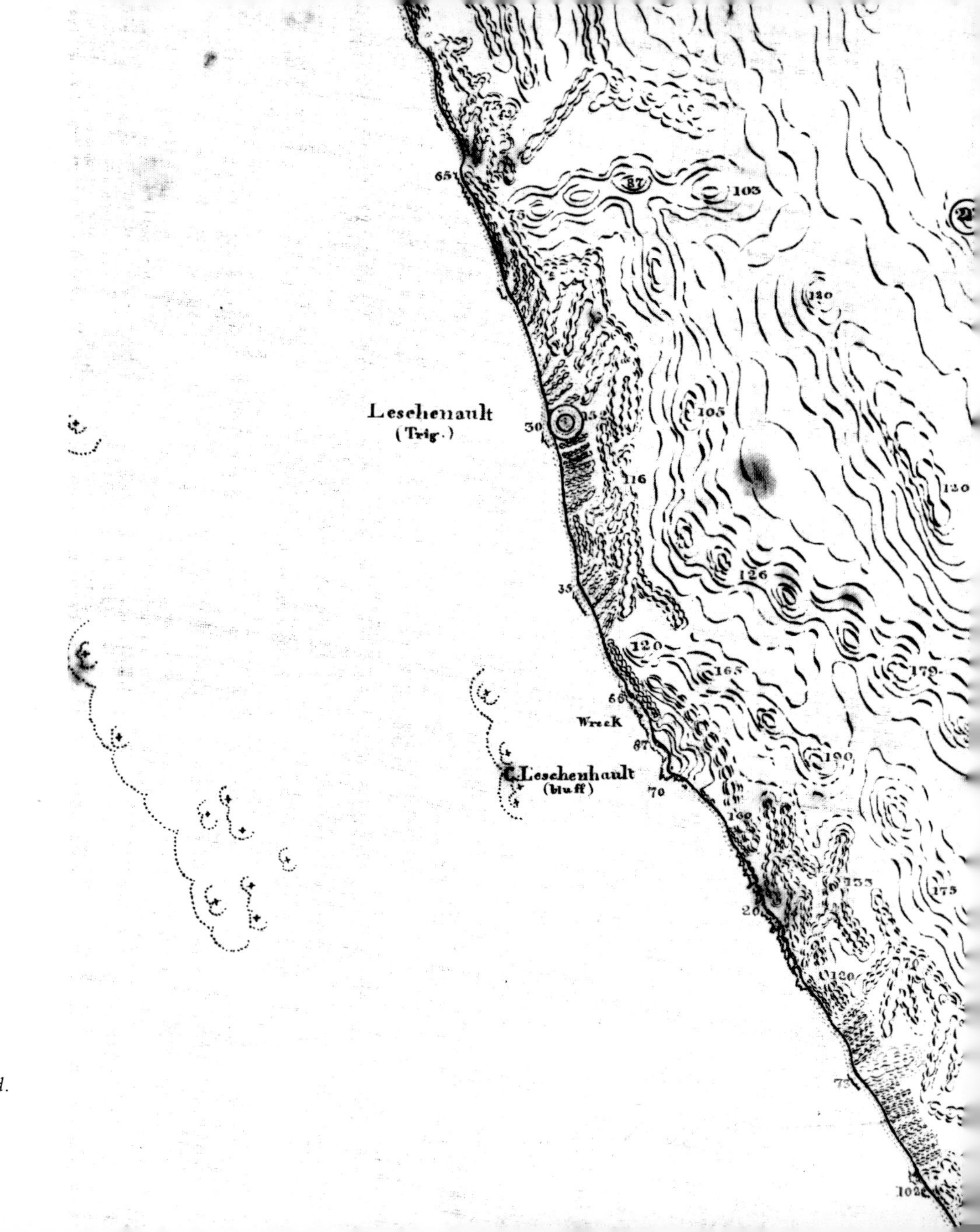

The wreck of the *Sea Bird*.

Bird ashore. The schooner struck the beach at 5 p.m. under a reefed mainsail, fore trysail and jib. (See map on p. 22.)

While the crew set about bringing cargo ashore, the passengers started to walk northward. They thought that they were not far south of Geraldton, but in fact the *Sea Bird* lay only 8 kilometres north of Moore River. The passengers were not the best of friends, and the arduous journey was marked by dissension. One of the men, John Shields, went lame soon after leaving the wreck. The others argued with him, but then left him with a blanket, some matches and some tea. Two of the men, Sullivan and Brown, left their companions and travelled together for 6 days, eating scrub and anything they could get hold of. On the seventh day, Brown saw Sullivan with a large knife, and as he knew Sullivan was famished, he feared that Sullivan would kill and eat him. He stole away and the two men arrived at Dongara separately.[2]

It was feared that John Shields might have perished where his companions left him, so several parties of police set out to find him. A patrol from Gingin found the wreck lying broadside on the beach 'inside a sunken reef and about 20 yards from the high bank and perfectly upright'.[3] The vessel was badly damaged, with a large hole in the port bow and another near her sternpost. The master and some of the crew were living in tents made from the *Sea Bird*'s sails.

Captain Hanham returned to Fremantle to face an inquiry. Attitudes in the Colony towards masters who lost their vessels had hardened in 1874. Hanham's certificate of competency was suspended for 2 years after the court found him guilty of negligence.[4] He had taken no sightings and neglected to keep a log book for dead reckoning calculations.

The wreck was auctioned at Gingin on 3 August. All the cargo had been saved in a damaged state. It included such miscellaneous items as coal, moleskin trousers, blue serge, shirts, yachting jackets, potatoes, pickles, raisins, currants, sardines, confectionery, oranges, stout, boots, cheese, and jarrah boards.[5]

The 40-ton *Sea Bird* (Official Number 36555) had two masts, one deck, a sharp stern and a wood frame. Her dimensions were 18.1 metres by 5.1 metres by 1.9 metres.[6]

NOTES

1. Charles Hanham, evidence at Preliminary Inquiry into the stranding of the *Sea Bird*, Fremantle, 20 July 1874, C.S.R. 786, fol. 62.
2. Report of W. Timperley, Geraldton, 22 June 1874, Police Records, Acc. No. 129, Battye Library.
3. Report of L. Back, Gingin, 27 June 1874, Police Records, Acc. No. 129, Battye Library.
4. Proceedings of Court of Inquiry, Fremantle, 20 July 1874, C.S.R. 786, fol. 62–75.
5. Lionel Samson, Auction of Wreck held at Gingin 3 August 1874, No. 118 in Auction Book, 29 December 1873-16 September 1879, pp. 73–6, Acc. No. 1120A, Battye Library.
6. Register of British Ships, Fremantle.

Contest

The barque *Contest* was built in Nova Scotia in 1860 and was one of the Black Diamond Line of Liverpool's fleet for a time.[1] In 1868, the vessel was bought by a Captain Simpson of Port Adelaide, and had her registry transferred to that port.

On 3 June 1874, the *Contest* arrived at Fremantle from Darwin, under the command of Captain Thomas Allen and carrying 50 tonnes of coal and seventy-five bags of copper ore.[2] After discharging, the vessel was piloted down to Rockingham, where she had been engaged to load railway sleepers for Lacepede Bay in South Australia. However, on 16 June, the barque was blown ashore on the beach at the Rockingham Timber Station. Later in the month, she was refloated, only to be driven back ashore, and on 14 July the Harbour Master reported:

> She is now lying with her head to the N.W., with her bows in 12 feet and her stern in about 2 feet, with the water as high inside her as out. She is very seriously hogged on her port side and strained greatly about the

Contest alongside the wharf at Port Adelaide in 1867.

> covering boards and deck, and as far as I am able to judge at present, the whole of her futtock timbers on the port side must be started from the floor timbers. Nothing is being done to her at present as the master expects instructions from the owners next mail.[3]

The vessel was condemned as a wreck and sold as she lay on the beach near the timber company's jetty. The new owner, Mr John Tapper, was ordered by the Harbour Master to remove the wreck from the beach within 10 days, under the provisions of 17 Vic. No. 4 Sec. 3, but it would appear that only part of it was taken away.[4]

In modern times, the remains of what seems to be the *Contest* wreck were found just 45 metres offshore adjacent to the Palm Beach boat ramp, in latitude 32° 16.5′ south, longitude 115° 42.8′ east, by members of the Maritime Archaeology Association of Western Australia. They were given a reward of $250. Association members later conducted a survey of the hull remains. The site is protected under the Historic Shipwrecks Act. (See map on p. 242.)

The 322-ton *Contest* (Official Number 37166) carried three masts and was built with a square stern, wooden frame, one deck with a half poop deck, and a female figurehead.[5] Her dimensions were 36.6 metres by 8.6 metres by 3.9 metres.

NOTES

1. *Inquirer*, 5 August 1874.
2. *Herald*, 6 June 1874.
3. Harbour Master to Col. Sec., 14 July 1874. Harbour Master's Letterbook, 1866–1883, Acc. No. 1056, Battye Library.
4. Harbour Master to Col. Sec., 6 August 1874, Harbour Master's Letterbook.
5. Register of British Ships, Port Adelaide.

Bertha

On the morning of 2 August 1874, the cutter *Fortescue* arrived in Port Walcott from the westward with the master, crew and passengers of the cutter *Bertha*. The *Bertha* had left Shark Bay on 16 July 1874 on a voyage to Port Walcott, but was wrecked on a reef off Point Cloates on 20 July.[1] (See map on p. 30.)

Although the vessel was lost with its entire contents, the five occupants, under the command of Joseph Moriah, escaped in a 5-metre dinghy. They left the *Bertha* on the night it was wrecked and proceeded towards Port Walcott, and after voyaging for 6 days without food or water, arrived at Tubridgi Point in Exmouth Gulf. Three local Aborigines prolonged some of the crew's lives by sharing around water and turtle meat. This help came too late for one of the men. When water was offered to 52-year-old pearler Charles Love, he tried without success to drink it, and died quietly that night.[2]

The four remaining survivors continued towards Port Walcott. Near Cape Preston, they met up with Captain Charles Tuckey in his cutter the *Hampton*. Tuckey refreshed the party and took them to Mardie Creek to meet the master of the cutter *Fortescue*.

When they finally reached Port Walcott Charles Love's 40-year-old widow Elizabeth was unable to eat or drink. She had been suffering for some months from dysentry, and died soon after being taken ashore.

The *Western Australian Times* described the *Bertha* as a pearling cutter owned by Fremantle chemist Daniel Congdon.[3] Sergeant Henry Vincent of the Roebourne Police stated that it was a 5-ton vessel.[4] It was not registered at Fremantle and may not have been registered at all. A 31-ton vessel of the same name (Official Number 61036), registered at Launceston and owned in 1871 by S. Griffiths and others, is listed in the Board of Trade Wreck Register of 1874 as having foundered at sea.[5]

NOTES

1. Sergeant R. Vincent, Roebourne, 4 July 1874, Report, Police Records, Acc. No. 129, Battye Library.
2. Report of Sergeant Vincent.

3. *Western Australian Times*, 4 September 1874.
4. Report of Sergeant Vincent.
5. F. C. Jarrett, *The Mercantile Navy List of Australia and New Zealand* (Sydney, 1871). See also Board of Trade Wreck Register, 1874, National Maritime Museum, England.

Enchantress

By the early 1870s, there were some eighty to a hundred small vessels pearling out of Cossack and other anchorages to the west. In 1872, Robert Sholl reported that there were thirty-one 'ships' and fifty-two 'boats' employed, and that the vessels ranged from 1 to 56 tons, the average being 10 tons.[1] The fifty-two 'boats' were dinghies. The smaller vessels could not stay at sea in heavy weather, and in 1873, the *Inquirer* saw the number of shipwrecks as suggesting:

> . . . the necessity of the early appointment of an official to see that the numerous coasting craft are fit to proceed to sea, and that, moreover, they are properly provisioned and in the charge of competent seamen. The recent lamentable loss of life connected with our pearl shell fishery, and the way in which the boats employed in it are almost without exception manned, calls for the prompt attention of the Government.[2]

Bigger 'ships' were, however, being despatched from the other pearling centres in Queensland and overseas. Two such vessels were the schooners *Flower of Yarrow* and *Enchantress*. Their owners were not content to restrict themselves to the well-known pearling grounds around Cossack.

In 1872, the Australian Fishing Company was floated in London, and these two yachts were fitted out in England to engage in the industry in the North West.[3] Ample capital was available. Lieutenant Ross, R.N. (perhaps one of the family from Melbourne who had been members of the abortive Camden Harbour Pastoral Association in 1863), was contracted for 3 years, and twenty-eight crewmen were said to have been specially chosen from the English fishing fleets.[4] In his prospectus, the promoter estimated that each diver could bring up 100 shells in an hour, an extreme exaggeration.

The 150-ton *Flower of Yarrow*, commanded by Lieutenant Ross, called at Fremantle, *en route* from London to the North-West, in March 1874. The *Flower of Yarrow*, which had formerly belonged to the Royal Yacht Squadron, carried two 21-pounder Armstrong guns and had on board a steam launch capable of accommodating twenty-eight passengers (a boon in heavy tidal waters).[5] She also carried equipment for conducting pearling operations on an extensive scale, and was said to have an auxiliary steam engine.[6]

The *Flower of Yarrow* sailed north to join the 29-year-old *Enchantress*, which came to the North West via Singapore, where she had called to recruit divers. The two vessels apparently commenced shelling as far north as Darwin, and gradually worked down the coast toward Camden Sound and King Sound.[7] They were finding abundant good-quality shell.

Lieutenant Ross would have been made aware by Roebourne Resident Magistrate Robert Sholl or others of the presence of shell around Brecknock Harbour. Sholl had spent some time there in 1864 after the wreck of the *Calliance*, and John McCourt had collected a small quantity of shell there in 1869 with the schooner *Argo*.[8] The *Enchantress* was filled with shell, and when the news of this success reached Roebourne, Sholl predicted that most of the larger vessels would leave Exmouth Gulf to try to link up with Lieutenant Ross' fleet.[9]

Disaster struck for the Australian Fishing Company when the *Enchantress* hit a reef off Champagny Island on 15 August. The vessel managed to reach Brecknock Harbour, where she became a complete wreck.[10] Then, in confrontations between the crew and the local Aborigines one European crewman, one Malay crewman and eight Aborigines were killed.

Lieutenant Ross had the *Enchantress* broken up to retrieve her copper fastenings.[11] But the pearling project was a failure, and the *Flower of Yarrow* was sold.

It seems likely that the *Enchantress* was beached at New Island, located in the entrance to Brecknock Harbour and having a freshwater spring. Several

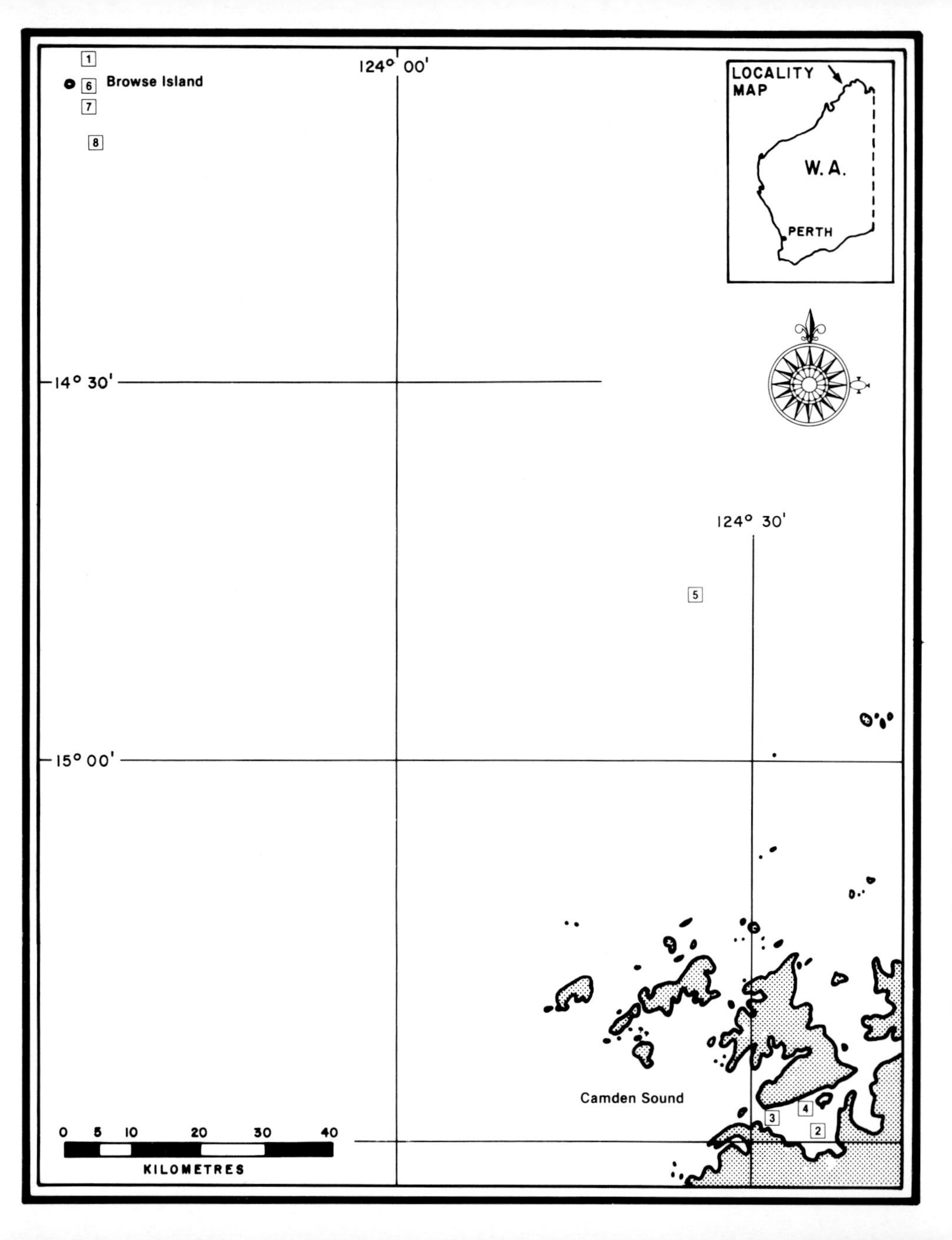
Browse Island
124° 00'
LOCALITY MAP
W. A.
PERTH
14° 30'
124° 30'
15° 00'
Camden Sound
0 5 10 20 30 40
KILOMETRES
1 Matterhorn
2 CALLIANCE
3 Enchantress
4 Unidentified Wreck 1838
5 Dutch Gunboat
6 RUNNYMEDE
7 Carleton
8 Sulina

expeditions from the Western Australian Museum have seen wreck material there. In 1963, Ian Crawford found heavy brass pipes, pieces of iron, broken china and a part of a compass[12], and in 1978, Scott Sledge found iron deck supports, bricks and lead piping.[13]

The schooner *Enchantress* (Official Number 25133) was built in 1846 at Cowes by White. The vessel was lengthened in 1860, making her dimensions 30.7 metres long by 7.2 metres broad by 3.4 metres deep.[14]

NOTES

1. Robert Sholl, Fisheries Report, January 1872, C.S.R. 714, fol. 17.
2. *Inquirer*, 11 June 1873.
3. Edwin Streeter, *Pearls and Pearling Life* (London, 1886), p. 160.
4. Mary Albertus Bain, *Full Fathom Five* (Perth, 1982), p. 74.
5. *Perth Gazette*, 20 March 1874.
6. Ron Parsons, The Pearling Fleet in Western Australia, in Gary Kerr (ed.), *Australian and New Zealand Sail Traders* (Blackwood, South Australia, 1974), p. 12.
7. *Inquirer*, 3 March 1875.
8. Robert Sholl to Col. Sec., Roebourne, 30 December 1869, C.S.R. 697, fol. 146.
9. Sholl to Col. Sec., Roebourne, 19 December 1874, C.S.R. 782, fol. 192.
10. Ron Parsons, *Australian Shipping Register*, January-February 1980.
11. *Inquirer*, 3 March 1875.
12. Ian Crawford, *The Art of the Wandjina* (Melbourne, 1968), p. 79.
13. Scott Sledge, *Report of Wreck Inspection North Coast, 1978*, Western Australian Museum, Perth, 1980.
14. *Lloyds Shipping Register*, 1874.

Grace Darling

The 1,042-ton United States ship *Grace Darling*, commanded by Captain A. P. Blevin, was stranded at Lockeville in Geographe Bay on the morning of 1 September 1874.[1]

At an inquiry into the stranding, Captain Blevin stated that he had come to Geographe Bay with the intention of loading jarrah sleepers under charter with the W.A. Timber Company.[2] He moored 5 kilometres from shore, but suffered from heavy weather from mid-July. On 11 August, he lost one of his anchors when the shackle-pin fell out. Although the Timber Company had divers on hand, they did not retrieve the anchor, because they were afraid of sharks. The *Grace Darling's* crew tied two cannon on the end of the cable where the shackle had parted, and used these as an anchor when the gales increased. However, the chain cable parted during a particularly strong gale and the vessel went ashore a kilometre north of the jetty.

The Captain was exonerated and the *Grace Darling* refloated. Nevertheless, the incident provided yet another stain on the reputation of Western Australian ports.

NOTES

1. *Inquirer*, 2 September 1874.
2. A. C. Blevin, evidence given to Preliminary Inquiry into the stranding of the *Grace Darling*, Lockeville, 14 September 1874, C.S.R. 786, fol. 77–98.

Cleopatra

The three-masted schooner *Cleopatra* was stranded on the east side of Pelsart Island, in the Southern Abrolhos, at about 4.30 a.m. on 12 November 1874. At the subsequent inquiry, one of the passengers, Maitland Brown, testified that he was roused by the whistling of the birds and went on deck, to see an island only 70 to 90 metres away.[1]

The *Cleopatra* was lightened of part of her cargo of lead ore and warped free on the high tide, before continuing on to Fremantle. Captain Edward Fothergill was censured by the court for failing to ensure a satisfactory lookout.

NOTE

1. Maitland Brown, evidence at Inquiry into the stranding of the *Cleopatra*, Fremantle, 20 November 1874, C.S.R. 786, fol. 104.

Cleopatra under full sail.

Centaur

The 25-year-old iron brig *Centaur* was wrecked about 24 kilometres north of Fremantle on 9 December 1874. The vessel had been on a voyage from Champion Bay.

At the subsequent inquiry, Captain Frederick Brabham stated that he had left Champion Bay on the 4th, laden with 200 tonnes of lead ore and four passengers, including the Surveyor General Malcolm Fraser and the prominent lawyer Septimus Burt.[1] Captain Brabham sighted Rottnest Island at 3.30 p.m. on 9 December, and he was standing in on the starboard tack heading towards the mainland when the vessel struck at 5 p.m. Seeing that nothing could be done to save the brig, Captain Brabham ordered out the boats, and within 25 minutes, all hands had left her. He estimated the position of the wreck to be 8 or 10 kilometres off shore.

However, the court employed Lieutenant Archdeacon to calculate the distance of the wreck from shore, and he came up with the much shorter distance of 2.4 kilometres. Captain Brabham was found solely to blame, having brought his vessel too close to the coast. His certificate was suspended for 6 months, but he petitioned the Governor and had this reduced to 3 months.

At the auction, held at Lionel Samson and Company's Cliff Street Fremantle store, the hull was knocked down for £46 and the lead ore for £80, while other items brought the total to £205.[2] Part of the lead cargo was salvaged.

The wreck was found in modern times by skindivers of the Blue Water Wanderers Club. Its position is latitude 31° 51.7′ south, longitude 115° 42.8′ east. The site is protected under the Historic Shipwrecks Act. (See map on p. 22.)

The 188-ton *Centaur* (Official Number 17568) was built at Aberdeen in 1849.[3] She was constructed of iron, with one deck, a square stern and a male bust figurehead. At the time of her loss, the vessel was owned by Mr A. J. Johnson of Melbourne.

A carved centaur figure at the Fremantle Museum. Of unknown origin, it is likely to have come from the wreck of the *Centaur*.

NOTES

1. Captain Frederick Brabham, evidence at Preliminary Court of Inquiry into the wreck of the *Centaur*, Fremantle, 11 December 1874, C.S.R. 786, fol. 115.
2. Lionel Samson, Auction Sale, 11 December 1874, Auction Book No. 129, p. 122, Acc. No. 1120/20, Battye Library.
3. *Mercantile Navy List*, 1874.

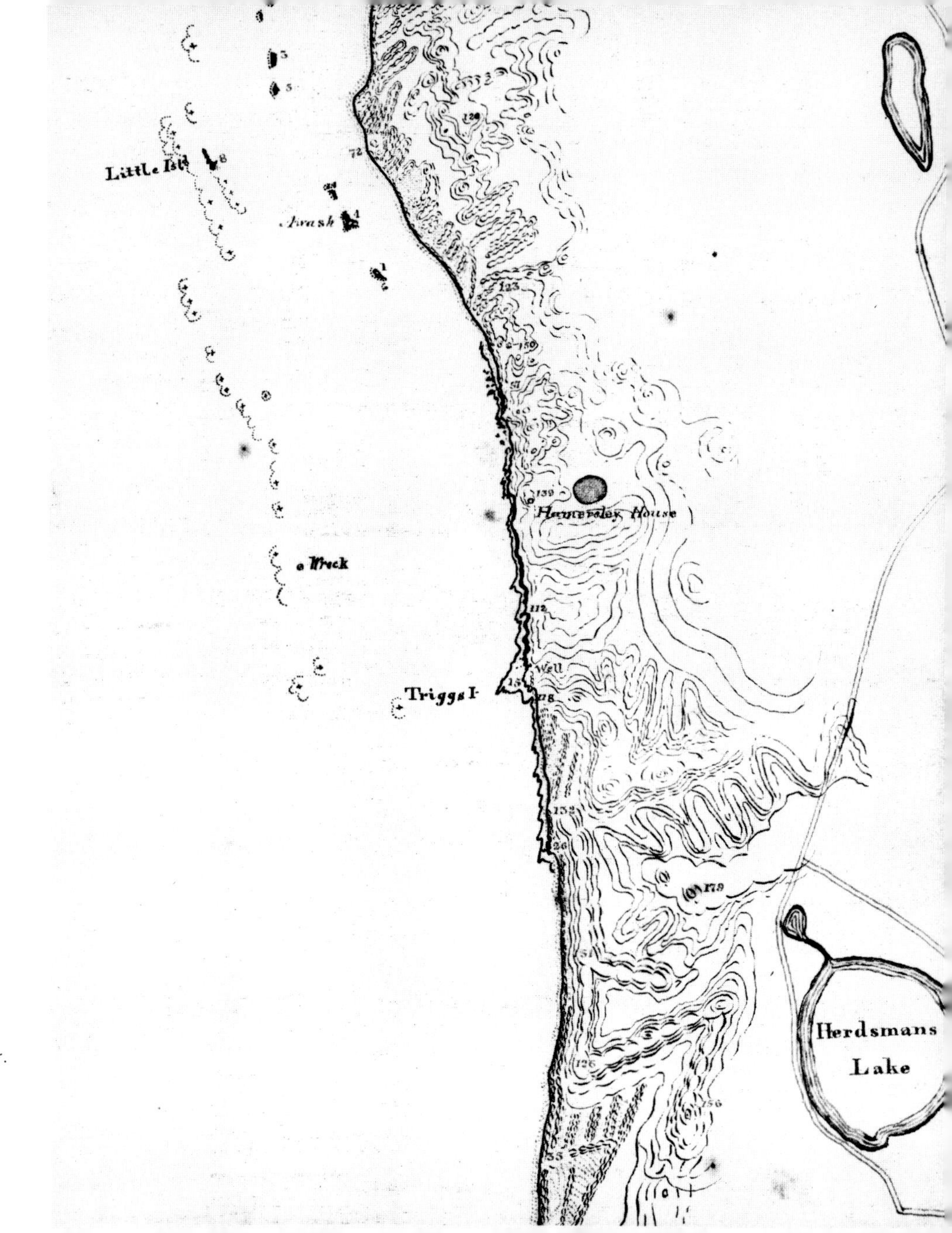

The wreck of the *Centaur*.

Blue Bell

Frank Goldsmith, in his *Treasure Lies Buried Here*, lists a 12-ton schooner named *Blue Bell* as wrecked at Broome in 1874.[1] He also lists a 12-ton schooner of the same name as having been wrecked at Broome in 1910. No supporting evidence has been found of Goldsmith's earlier listing, which seems dubious. A 65-ton schooner of the same name (Official Number 32431), registered at Sydney, foundered at Albany Pass, near Cape York, on 7 May 1874.[2]

NOTES

1. Frank Goldsmith, *Treasure Lies Buried Here* (Perth, 1946), p. 203.
2. Register of British Ships, Sydney.

Zedora

Fremantle Harbour Master George Forsyth was called out on the morning of 11 February 1875 by the Assistant Pilot, who reported a vessel on the reef near the Mewstone. Forsyth later stated:

> I immediately started in company with Water Police to the vessel. I found her to be a small barque laying on a clump of rocks about 1¼ miles to the W. of the Mewstone. Found the vessel had run on the rocks with her bow on them and careening over to port with her starboard bilge out of the water and heading to the westward.[1]

The barque *Zedora* had left Mauritius, bound for Adelaide, in sand ballast, under the command of John Hodges, with a crew of ten. On Sunday, 7 February, when the vessel was in latitude 37° south, a strong squall struck and shifted the ballast. The *Zedora* was thrown on her beam ends and remained in that precarious situation for about 2 hours, during which time the crew worked down below, throwing the ballast back in position. The vessel was righted and answered to the helm, but the next morning when the hatches were lifted, the crew found that the pumps were choked with sand, and water was washing about in the ballast.

Captain Hodges changed course for Fremantle, intending to seek a cargo there. At noon on the 10th, the *Zedora* was 145 kilometres west of Rottnest. Hodges had the crew furl the courses and spanker, so that no sail was in the view of the horizon, and steered south-easterly at 14 kilometres per hour, expecting to sight the Rottnest light at about 8 p.m. But the Rottnest light was either obscured by haze or not working, so the *Zedora* sailed past the island and through the Challenger Passage.[2]

Hodges went on deck at 11 p.m., and at about the same time, the lookout saw land ahead. The course was immediately changed to west-north-west, but at 12 p.m., the vessel struck a reef and became unmanageable. Just as the long-boat was launched, the *Zedora* floated free.

Things were still looking very grim on board. The pumps were choked, the distress rockets would not work, and the anchors were not ready. They had not been catted, because the moving sand ballast made the vessel unstable, so the *Zedora* had to keep going. The course was set at west by south, and the vessel was doing 5 kilometres per hour when the lookout reported a revolving light on the starboard bow (the Rottnest light), bearing north-west. At the same time, broken water was seen ahead.[3]

Captain Hodges later described the ensuing chaos:

> I ordered the mate to cast the lead and in so doing the lead carried away [Hodges himself had lost another sounding lead earlier in the evening], and immediately the vessel struck aft carrying away the rudder and the wheel (which I was at at the time) and began to bump very heavily. I ordered the gig to be again lowered at once, the man on the long-boat singing out that she was sinking on the rock astern. The vessel then went over on her side, and settled down at once.[4]

All the hands tumbled into the boats, saving nothing but what they were wearing, and hurriedly sailed away from the wreck to avoid being struck by falling spars. They reached the mainland 10 kilometres north of Fremantle at 8 o'clock in the morning.

A preliminary inquiry assembled at Fremantle by the Collector of Customs exonerated the Captain, finding:

> We attribute the accident entirely to Rottnest light not having been seen from the vessel, when she must have been within the usual range of that light for two hours prior to the casualty, though a proper lookout was kept.[5]

This was not good enough for the Governor, who felt that the court's opinion was equivalent to grave charges against the lightkeeper at Rottnest.[6] He wanted the blame to be laid with the Captain, rather than with the Colony's navigational facilities. So the issue went back to court, with formal charges being laid against Captain Hodges. But Hodges had found several reliable witnesses who had been on sentry duty at Fremantle on the night of the wreck, and they had noticed the Rottnest light by its absence. Thus, the second court concurred with the first in its findings.

The hull of the *Zedora* was sold at auction for £160, but the sale of fittings brought the total to over £400, so it is certain that a good deal of material was salvaged from the wreck at that time.[7] A year later, parts of the wreck were found washed ashore as far away as Jurien Bay, over 100 kilometres north of Fremantle.[8]

The site of the *Zedora*, lying at latitude 32° 04.1′ south, longitude 115° 37.7′ east, is well known to modern divers. During the 1960s, the wreck was given a lot of attention by eager treasure hunters, who wrongly thought that it was the *Lancier* (see *Unfinished Voyages 1622–1850*). Little remains on the *Zedora* site today, apart from some heavy timbers. It is protected under the Historic Shipwrecks Act. (See map on p. 242.)

The 269-ton barque *Zedora* (Official Number 62891) was built in 1869 at Bideford in Devon by Johnson, and was owned by J. Mill. The vessel was 35.9 metres long by 7.6 metres broad by 4.5 metres deep, built of oak and classified A1.[9]

NOTES

1. George Forsyth, evidence at Inquiry into the stranding of the *Zedora*, Fremantle, 13 February 1875, C.S.R. 813, fol. 12.
2. A later report stated that the vessel actually went into Owens Anchorage before turning. See evidence of Thomas Shaw in Royal Commission into Shipping Rings, *Votes and Proceedings*, 1905, 2nd Sess., p. 101.
3. John Hodges, evidence at Inquiry.
4. John Hodges, evidence.
5. Preliminary Inquiry, C.S.R. 813, fol. 21.
6. Governor to Collector of Customs, 12 February 1875, C.S.R. 56, fol. 89.
7. Lionel Samson, Sale, 12 February 1875, Nos 137 and 138, Auction Book, Acc. No. 1120A, Battye Library.
8. *Western Australian Times*, 20 June 1876.
9. *Lloyds Shipping Register*, 1874.

Boat, Cape Naturaliste, 1875

Two young sailors, who had deserted from timber ships, took a boat from alongside the jetty at the Vasse on 22 February 1875 and went to sea. They had been drinking a good deal beforehand, and after they left the wind blew very hard from the south-east, so they were given up for lost. The next evening, however, they reported themselves at the police station and said that they had beached the boat near Cape Naturaliste, and walked back. It was later reported that the boat had become a wreck.[1] (See map on p. 204.)

NOTE

1. *Herald*, 27 February 1875.

Geffrard

The brig *Geffrard* was built on the English Channel Island of Jersey in 1853 by Fred Clark. In 1872, the vessel was sold to Melbourne master mariner Fred Davis, who employed her carrying cargo to ports between Melbourne and Shanghai.[1] On these voyages, the *Geffrard* would sometimes call at Fremantle and Champion Bay, seeking cargo.

During one such voyage in 1874, the *Geffrard* left Fremantle, in ballast, bound for Quindalup in Geographe Bay, where cargoes of timber were available. A similar route was followed the next year, the vessel departing Fremantle on 19 April.

By 12 June, the *Geffrard* had on board a full load of timber for Adelaide. When Captain Munday went ashore in the evening to complete his business with Henry Yelverton, the weather was calm, but at 6 a.m. on the 13th a fresh northerly breeze sprang up, increasing to a gale during the next 4 hours.[2]

The *Geffrard* was lying half a kilometre off the beach, about 1.25 kilometres to the east of Yelverton's timber station jetty. The chief mate, George Allen, described the events of the afternoon:

> Weather continued same till 7 p.m. when it fell calm with heavy rain for about 20 minutes, when the wind came suddenly with hurricane violence. At 7.30 chain parted. Immediately let go starboard bower, paid out to about 45 fathoms. When it parted ship canted to the north-west. Hoisted the fore-top-mast staysail and set main lower top sail. Ship struck aft first on account of drawing 2 feet more than forward. Kept the top sail set. 8.30 sounded the well and found 9 inches. Ship lying on the port bilge not striking heavily.[3]

The *Geffrard* continued to move towards the shore, and finished up on the beach 2.5 kilometres to the east of the jetty. There, the vessel remained. In July, Lloyd's agent, Wallace Bickley, called tenders for unloading the cargo, and the following month, the wreck and the cargo were auctioned, the hull going for £200 to Fremantle merchant Elias Solomon and the cargo at £413 to Fremantle butcher and shipowner William Pearse.[4] Captain Munday was exonerated by the court of inquiry. (See map on p. 204.)

The 316-ton *Geffrard* (Official Number 23252) had one deck, a square stern, a wooden frame and a male figurehead. Her dimensions were 37.2 metres by 7.3 metres by 5 metres.[5]

NOTES

1. Register of British Ships, Melbourne.
2. J. W. Munday, evidence at Court of Inquiry into the casualty to the brig *Geffrard*, Busselton, 7 July 1875, C.S.R. 813, fol. 212.
3. George Allen, evidence at Court of Inquiry.
4. *Inquirer*, 11 August 1875.
5. Register of British Ships, Melbourne.

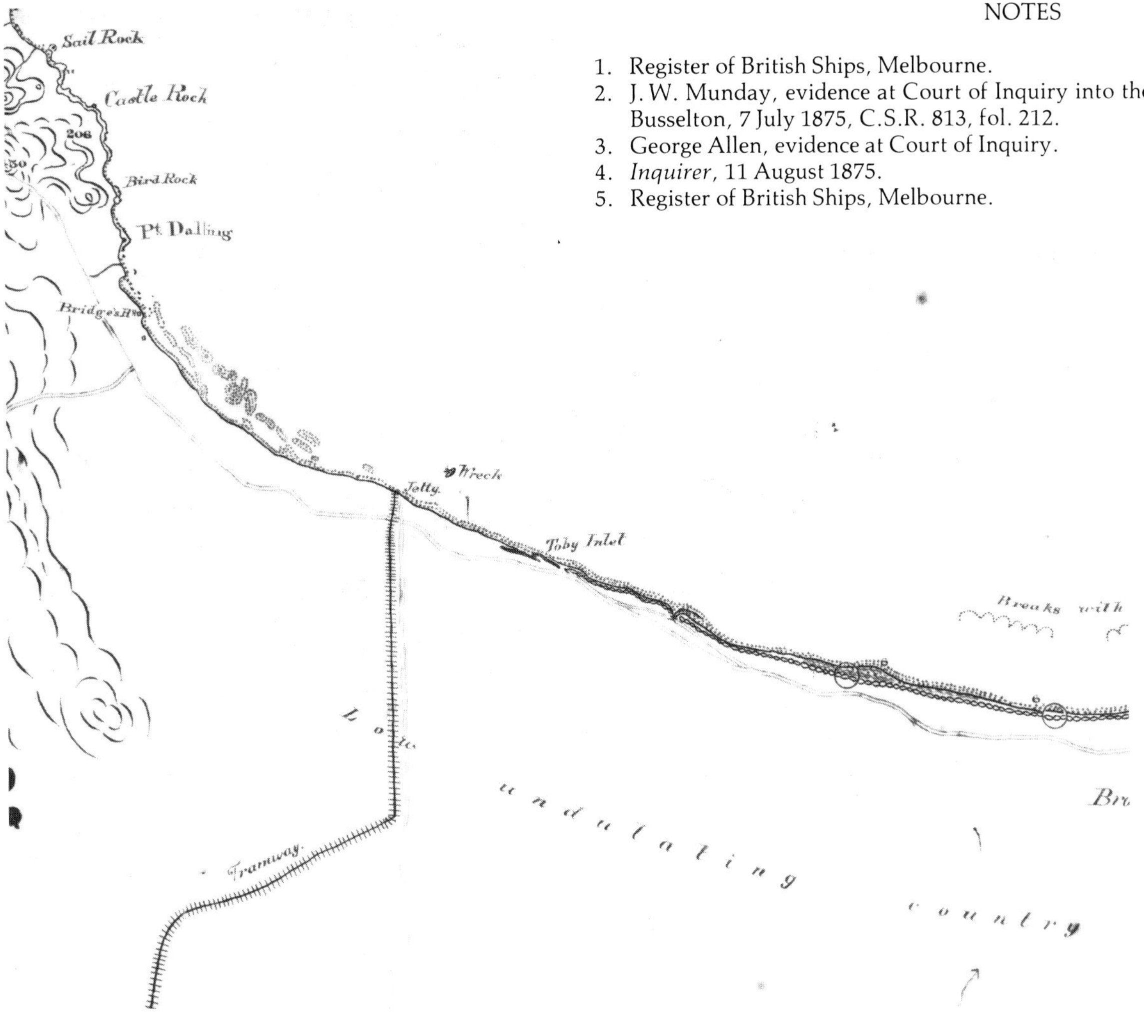

The wreck of the *Geffrard*.

Victory

The cutter *Victory* departed Albany on 26 June 1875, bound for Fremantle. The vessel was driven ashore by bad weather at West Cape Howe, and wrecked.[1] The captain and crew were saved.

The 24-ton *Victory* (Official Number 61119) was built at Albany in 1873, with a wood frame, one deck, one mast and an elliptical stern.[2] Her dimensions were 14 metres by 4.3 metres by 2 metres. She was owned by Captain William Douglas, of Albany.

NOTES

1. *Herald*, 3 July 1875.
2. Register of British Ships, Fremantle.

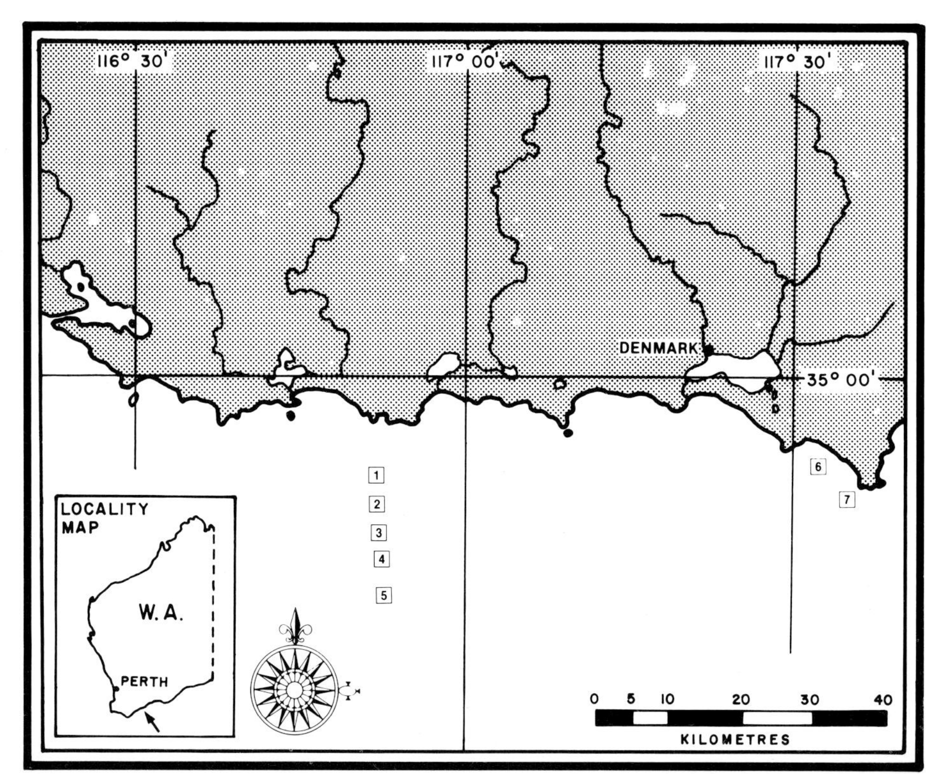

1 Unidentified Wreck 1840
2 *Mary Herbert*
3 *Hope*
4 *Brothers*
5 *Knowsley Hall*
6 *Harlequin*
7 *Victory*

Boat from Subahani, 1875

Aboriginal prisoners on Rottnest Island found a wrecked boat on 5 July 1875. The planks were dowelled together, suggesting to observers that the boat was a 'sampan'. Pilot Henry O'Grady examined the material, on the north-west side of the island, and recognized the boat as one that had been lost from the 86-ton pearling schooner *Subahani*.[1] The schooner itself was not lost at the time. (See map on p. 242.)

NOTE

1. H. O'Grady to W. Clifton, 13 July 1875, C.S.R. 795, fol. 72.

Alert

On 28 July 1875, the electric telegraph sent the news from Pinjarra to Perth that the cutter *Alert*, bound for Fremantle, had gone ashore on the Murray River bar, and looked like becoming a total wreck.[1] (See map on p. 242.)

John Kelly was the master of the vessel, but he held no certificate. He told the court of preliminary inquiry:

> I proceeded on 22 July to the Murray for the purpose of loading sandalwood inside the bar at the Murray jetty. Went inside on Sunday morning, discharged cargo and loaded. On Wednesday the 28th proceeded to sea, started with wind S.S.E. and tried to get out over Murray bar. A heavy sea on her, struck the ground, strong current setting out at the time. Struck very hard aft, shipped seas all over, and slewed broadside to the sea, and in about one hour afterwards she filled and has since become a total wreck. Before taking the bar I sounded it and found 6 feet of water, the boat only drawing 4 feet 8 inches. A fair wind out.[2]

Kelly knew that it was a dangerous bar in winter, but the owners had directed him to make the crossing. The court found that Kelly was to blame, but as he was not certificated, they had no power to hold a formal inquiry. The *Alert* was not insured.

The 19-ton *Alert* (Official Number 52239) was built at Fremantle in 1867 by J. C. Mews, and was sold to Edward Newman.[3] The vessel had a wood frame, one deck, and a square stern. Her dimensions were 14.4 metres by 4 metres by 1.6 metres.[4]

NOTES

1. *Western Australian Times*, 30 July 1875.
2. John Kelly, evidence at Preliminary Inquiry into stranding of *Alert*, Fremantle, 17 August 1875, C.S.R. 795, fol. 91–2.
3. *Mercantile Navy List*, 1874.
4. Register of British Ships, Fremantle.

James Vinicombe

The Rottnest Pilot, Henry O'Grady, picked up a lifebuoy and a green-painted water cask on the south-east side of the island in August 1875.[1] Both items had been in the water a long time. The lifebuoy bore the name *James Vinicombe*.

A 638-ton barque of that name was built in Sunderland in 1859, and owned by William Adamson, of Sunderland.[2] It was last entered in Lloyds in 1879, and never visited Western Australian ports. It seems likely that the lifebuoy drifted in from thousands of miles out to sea, and that it did not represent a shipwreck.

NOTES

1. H. O'Grady to G. Forsyth, 3 August 1875, C.S.R. 800, fol. 131.
2. *Mercantile Navy List*, 1875.

Dania

In September 1875, the cutter *Dania*, on a coasting voyage from Bunbury to Fremantle, struck a rock in the South Passage and had to be run ashore on the beach about a kilometre north of Rockingham.[1] The master, Hugo Leidicke, discharged his cargo of potatoes and butter. In October, the vessel was expected to become a complete wreck,[2] but it was repaired and sailed for Bunbury on 23 December.[3]

The *Dania* survived for many more years, although not without incident. In January 1883, the vessel piled up in the Yammadery Creek, along with another craft being used for pearling. The *Dania* was salvaged and brought back to Fremantle, where Thomas Mews lengthened the hull and converted her to schooner rig, with the idea of employing the vessel for freighting grain from Dongara.[4] It seems that the cutter was re-registered and renamed *Alma* at this time, although the Register states that the *Dania* was broken up in 1886.[5]

The *Dania* (Official Number 61120) was built at the Vasse in 1874 as a 25-ton round-stern cutter, with the dimensions 13.3 metres by 4.3 metres by 2.3 metres.[6] Her first owners were Thomas Johnson (a Vasse mariner) and George Shenton.

NOTES

1. V. Fall, *The Sea and the Forest*, p. 70. See also *Inquirer*, 15 September 1875.
2. *Western Australian Times*, 8 October 1875.
3. *Western Australian Times*, 24 December 1875.
4. *Inquirer*, 22 April 1885.
5. *Inquirer*, 3 March 1886.
6. Register of British Ships, Fremantle.

Mary Herbert

Unfortunate events surrounding the building of a ship sometimes give the vessel a reputation of being jinxed. The *Mary Herbert*, one of the larger vessels built in the Colony, acquired such a reputation. The short career of the vessel, built and lost in 1875, was remembered by W. Lynch in a letter to the editor of the *West Australian* in 1941:

> The first boat I saw built and launched was the schooner *Mary Herbert*. She was built by an old shipwright by the name of Jackson, and he built her for a publican named Herbert. When the boat was about half built he had a dispute with the owner, and he flung in his job and for about 8 years she remained in that state. Then, with his 3 sons, he again took on the job and finished her, but the old man did not launch her. There was again a dispute between Jackson and Herbert, and his leading hand, Fred Jones, did the launching, or tried to. The old man predicted she would never see the water. He was not quite right, but near enough; she broke down in the slipway. Just as the owner's wife broke a bottle of wine over her, and called her the *Mary Herbert*, after herself. She had just then got to the water's edge and remained there for weeks before they got her into deep water, when she was rigged and completed and got ready for her first voyage, which was to be Hobart. Her first passengers were to be Lady Weld, the Governor's wife, and her two daughters; also the owner, Mr Herbert, who was making the round trip in her. The boat made a quick passage to Hobart, and on her return trip picked up a load of flour at Adelaide for Fremantle. She called in at Albany, and left that port for Fremantle (early in September), but neither she, her owner, or captain and crew were ever heard of again.[1]

In Fremantle, it was speculated that the *Mary Herbert* had either foundered in a storm or been wrecked on the coast between Albany and Augusta. One search party found a panelled locker door between the mouths of the Donnelly and Warren Rivers, and a lifebuoy near the Gardiner River.[2] A branded log of sandalwood, supposed to be part of the *Mary Herbert* 's cargo, was picked up 20 miles west of Albany.[3] But no other material was found at the time, so it seems probable that the vessel went down in deep water, which may suggest that her disgruntled builder had not produced a sound craft. (See map on p. 164.)

The *Mary Herbert* was insured, but the call at Albany was not included on her bills of lading, and the insurers denied liability because the ship had deviated from her stated voyage. Claimants were directed to seek recompense from Mr Herbert's estate.[4]

The hope of finding clues as to the fate of the *Mary Herbert* made local residents take an interest in a variety of flotsam in the area. In 1876, initial news of the finding of a note in a bottle led to some speculation of another wreck. The topsail yard of a schooner had been found near the mouth of the Warren, and the police constable who followed up the search in September 1876 found a corked bottle containing a note, which read:

> Ship *Ethiopian*. In latitude 44° S longitude 42° E from London bound to Sydney N.S.W., 57 days from the Lizard and distance sailed 9,635 miles, average per day 169 miles. All well on board. 6 June 1875 William Faulding.[5]

The yard would not have been from the *Ethiopian*, which was still afloat some years later.

The 92-ton brigantine *Mary Herbert* (Official Number 72471) was built at Fremantle in 1875, with a wooden frame, female figurehead, one deck and an oval stern. Her dimensions were 23.3 metres by 6.3 metres by 2.5 metres.[6]

NOTES

1. W. Lynch, letter to the Editor, *West Australian*, 10 April 1941.
2. *Inquirer*, 17 November 1875.
3. *Western Australian Times*, 10 March 1876.
4. E. M. Ashmore to W. Duffield, 2 May 1876, Samson Collection, Battye Library.
5. Note attached to report by P.C. L. Stock, Bunbury, 17 September 1876, Police Reports, Acc. No. 129, Battye Library.
6. Register of British Ships, Fremantle.

Fairy Queen

The schooner *Fairy Queen* sailed from Singapore on 3 August 1875 on a pearling cruise to the North West coast of Western Australia. The vessel was owned by Messrs Marmion, Brown and Gill. The pastoralist and pearler William Marmion was the managing owner. Captain Andrew Edgar joined the *Fairy Queen* as master in Singapore. On leaving port, she had on board a crew of thirty-eight, including divers and, amazingly, one stowaway.[1]

Captain Edgar took the *Fairy Queen* down the north-east side of Sumatra and turned south through the Sunda Strait, where it began to blow. On the night of 10 September, the tiller broke away and split the rudder head. The crew fitted a jury tiller and stood the vessel back on course. The *Fairy Queen* was standing across the trade winds and experienced strong breezes and squalls.

In latitude 27° south, strong gales carried away the deadeye of the fore topmast backstay. Captain Edgar intended to make the land and go into Shark Bay. But although he sighted Cape Cuvier at 11 a.m. on 7 October, he could not get further south against the south-westerly winds, and decided to seek shelter in Exmouth Gulf. He rounded North West Cape just as it was getting dark, intending to shelter under the lee of the Cape. But it was too rough to stay there, so he reduced sail and stood towards Y Island, some 27 kilometres to the south-east. When he reckoned himself to be close to the island, he furled the foresail and put the vessel on a port tack back to the west. He was recalled to the deck by the mate, Andrew Heather, at 3.45 a.m. the following morning:

> I asked him what was the last sounding he got, he told me 12 fathoms. I told him to call the hands up to wear ship. I could not see the land then. It was very dark and blowing hard. I took the wheel myself. The mate and the savant [a wise man in an Asian crew] were on the forecastle with two or three hands working the head sheets. The man I relieved from the wheel I sent to prise off the main sheet. I told them to haul the head sheets to windward. I found that she would not go off and called out to let go the peak halyards. The man that let go the peak halyards I called to haul in the mainsheets as the sail was in the water. She still kept coming up to the wind and would not pay off. The mate sang out 'what is the matter that

> she won't come round?' I put my hand down and found that the wheel ropes had slipped off the barrel and were hanging loose on the spindle. While I was trying to slip the barrel forward the mate sung out 'close to land'. I called out 'let go the anchor'. We let go the anchor. In a few minutes afterwards she struck on the starboard bilge.[2]

Five minutes later, the vessel was stuck fast, and the seas were breaking over the port side. At daylight, the deck began to open up. The crew could see the low outline of the shore close by. Divers walked out the anchor underwater, while others took goods ashore to lighten the vessel. But during the day, the *Fairy Queen* began to break up. At midday on the 12th, short of water, the crew launched the *Fairy Queen's* boats on the lee side and headed east towards the pearling grounds. About 5 kilometres from Y Island, the mate's boat shipped a sea and capsized, but the other boats rallied to its aid and bailed it out. They reached the Mary Ann Patch 2 days later and found the cutter *Swan*, owned by Brown and Gill. This vessel took Edgar and some of his crew in to Tien Tsin on the 18th. (See map on p. 186.)

The *Fairy Queen*'s loss had an unhappy sequel. Two of the owners, Brown and Gill, had engaged a number of Gascoyne Aborigines at Shark Bay for pearl diving. These men were on the *Swan* when the *Fairy Queen* survivors arrived. To make way for the *Fairy Queen* men on the *Swan*, nine of the Aborigines were transferred to another cutter, the *Albert*, making that vessel overcrowded.

The Government Resident at Roebourne later gave his version of the subsequent events on board the *Albert*:

> On the day in question the natives on board the *Albert* rushed aft. One of them named Nirba struck W. [probably William] Ross, at the time in charge of the boat, on the left temple with a weero [woomera, or spear thrower?]. He ran below for arms and found an unloaded musket with which he returned on deck. When Ross ran below Nirba attacked the man at the wheel, who scuffled with him. The man's name is Andrew Heather, late mate of the *Fairy Queen*. Another native named Ballemerda attacked a third white man named Charles Holkensen, lately a seaman on board the *Fairy Queen*, with a piece of oar. He took the weapon from him, upon

which the native got hold of a spy glass, which was on the companion, and smashed it upon Holkensen's head, giving him a severe wound. At this time, while the men were struggling, Ross came on deck and struck first Nirba and then Ballemerda on the head with the butt end of the musket, and the stock breaking, struck the last named native, who was threatening him, with the barrel. Ross also hit another of the Gascoyne natives who advanced toward him with a knife. This last man and four others jumped overboard, the *Albert* being half a mile from the shore. The two natives Ballemerda and Nirba threw themselves into the hold, whither another of the natives named Wilga had previously rushed and concealed himself under some firewood. The hatches were fastened down and the *Albert* beat up for the pearling fleet. At sundown the natives were taken from the hold and secured with canvas gaskets.[3]

Diver Colin Powell brings a small cannon ashore from the *Fairy Queen* wreck.

Nirba and Ballemerda, who had been caught in the hold, both died in captivity soon afterwards, and the Government Resident was given the task of determining the cause of their deaths. He decided that Ross had acted in self-defence, but concluded that the attack was made because of the vessel's proximity to the shore, the Aborigines believing that by killing the white men, they could get back to their country unmolested.[4] He also observed:

> It seems that the natives were not used to diving and I think it was unwise to engage them as divers, for useless men on board ship do not make friends.[5]

Meanwhile, the wreck of the *Fairy Queen* was sold at auction, and Captain Edgar was called upon to face an inquiry. The court found that Edgar had insufficient knowledge of the tides and currents in the Gulf, but in the absence of reliable charts of the area, they were not prepared to suspend his certificate.

In modern times, the wreck was found by T. Coleman and J. Labnowski and reported to the Western Australian Museum. The finders were given a reward of £100 by the Trustees of the Museum. The wreck lies about 10 metres from shore, 100 metres south of the U.S. Navy Communication Station's pier at North West Cape, in latitude 21° 49.2′ south, longitude 113° 11.5′ east. When Museum divers inspected the site in 1973, visible material included a small cannon, several anchors, and some ballast bricks and stones. The cannon and one anchor were raised and treated in the Museum's Conservation Laboratory.

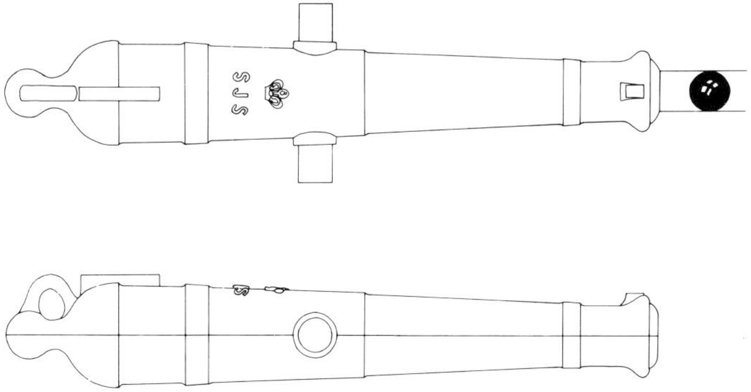

Drawing of the gun after cleaning.

The cannon may have had any one of several functions on board the *Fairy Queen*. The schooner would, if she had arrived safely at the pearling grounds, have acted as a mother ship, sending a number of small boats around the pearling grounds during each day. Signalling devices would have been of great importance in such circumstances. The *Fairy Queen* gun had initially been designed for defensive purposes rather than signalling, because it is properly disparted.[6] The *Fairy Queen* had been in the waters around Singapore, where defence against pirates was sometimes needed.

Given the poor relations between employer and 'employed' in the pearling industry at that time, the gun may also have had the effect of intimidating Aboriginal divers. Aborigines were kidnapped by armed pearlers from their tribal grounds and treated very much as slaves. On occasion, they were killed for refusing to dive. Others were sold with the luggers on which they were employed.[7] The prisons in towns like Carnarvon and Roebourne were described as hardly more than stockyards where humans were broken in preparation for employment. Islands near North West Cape were used as depots for kidnapped Aborigines, and small cannon kept on pearling vessels such as the *Fairy Queen* would doubtless have made it easier for the pearling masters to suppress their divers.

The piece is a 1.37-metre-long 2-pounder, made of iron and bearing the letters SJS with a crown. Brigadier O. F. G. Hogg has suggested that these letters are derived from the Portuguese Order of St James of the Sword, which was introduced into Portugal from Spain in 1290.[8] It is possible that the gun could have been bought from an arms dealer in Singapore or from Portuguese sources in Timor. Estimates of the date of manufacture of the gun range from 1790 to 1860.[9]

The 115-ton *Fairy Queen* (Official Number 71529) was built as the Dutch schooner *Rhio* in Singapore.[10] It apparently changed its name and rig when bought by Marmion and others in July 1875. Given the number of equipment failures on her final voyage, it may be assumed that the vessel was old and poorly maintained.

NOTES

1. Andrew Edgar, evidence at Court of Inquiry into the wreck of the *Fairy Queen*, Cossack, 8 November 1875, C.S.R. 809, fol. 154–158.
2. Andrew Edgar, evidence at Court of Inquiry.
3. Government Resident (Roebourne) to Col. Sec., 4 November 1875, C.S.R. 809, fol. 174–176.
4. Government Resident to Col. Sec.

5. Government Resident to Col. Sec.
6. H. L. Blackmore (Assistant Keeper, The Armouries, H.M. Tower of London) to G. Henderson, 19 September 1972.
7. Su-Jane Hunt, The Gribble Affair, pp. 41, 56, 94.
8. M. L. Pearl (The Metals Society, London) to G. Henderson, 7 February 1975.
9. Blackmore and Pearl. Mendel Peterson (Smithsonian Institution) to G. Henderson, 6 February 1973.
10. Board of Trade Wreck Register, 1875. See also *Mercantile Navy List*, 1876.

Stefano

When the brig *Alexandra* arrived at Roebourne in May 1876, the master, George Vinal, reported that he had passed the pearling cutter *Jessy*, 65 kilometres north-west of North West Cape and bound for Fremantle. There had been too much wind and sea for spoken communication, but as the two vessels approached each other, the men on the *Jessy* had held up a board chalked with the words:

> Barque *Stefano* wrecked on NW Cape. I have the only two survivors on board.[1]

Captain Vinal approached North West Cape to search for wreckage and survivors, but gesturing Aborigines appeared on the shore. The *Alexandra*'s men feared that the Aborigines were trying to lure them in to the beach for an attack, so they left for Roebourne to break the news of the wreck.

When the *Jessy* reached Fremantle, the two shipwreck survivors, 16-year-old midshipman Michele Baccich and 19-year-old crewman Ivan Juric, told a sad story of shipwreck, privation and deliverance from death.

The Austrian barque *Stefano* had sailed from Cardiff in Wales, bound for Hong Kong, on 31 July 1875 with 1,300 tonnes of coal and a complement of seventeen. The vessel rounded the Cape of Good Hope and steered for the west coast of Australia, where Captain Vlaho Miloslavic intended to check his longitude. Land was sighted in the vicinity of Cape Cuvier on 26 October, but the vessel was wrecked further north at 2 a.m. the following morning. Baccich described the events:

> It was to the southward of Point Cloates on an outlying reef about six miles from the shore. The vessel was sailing with all sails set going at the rate of 9 knots, and ran on the reef without warning steering by compass N by E. Directly she struck the sea swept the deck and the crew took to the masts to wait for daylight. About 2 hours after she struck the masts fell overboard. I, the Captain and mate and an AB managed to get into the gig.[2]

The gig immediately capsized in the breakers, but Baccich survived by clinging to the keel of the upturned boat, and after 10 hours in the water, he was washed ashore. The *Stefano* broke up soon after Baccich's departure, but eight others reached the shore in the same area by clinging to parts of the wreck. (See map on p. 30.)

The young sailors thought themselves to be not far north of the Gascoyne River, so, after burying Pavo Radovich, whose body had washed ashore, and establishing a camp, they gathered provisions together with the intention of walking southward along the coast.

A small group of Aborigines walked into the camp on the 31st. They were unable to tell the sailors where they were, because the two groups lacked a common language; the Aborigines could speak English, but the crewmen could not. The Aborigines had on previous occasions had contact with the numerous pearlers who operated in Exmouth Gulf and even at such places as Yardie Creek, on the west side of the Cape. They brought to the camp a chart, which had washed ashore from the wreck, showing the North West Cape area and the coast to the south.

On 1 November, the little band of survivors, no wiser from examining the chart, set out on the journey southward.[3] Seven days later, they had reached Cape Farquhar, where they were given water by a group of Aborigines, and were reunited with a tenth member of the crew, Guiseppe Perancic, who had come ashore on a lifeboat some 16 kilometres south of the wreck.

The sailors left the Aborigines and continued on to Cape Cuvier. Desolate country south of Cuvier forced them to turn around on 16 November, and 4 days later, they were back at a water-hole in the vicinity of Cape Farquhar, where they subsisted on rock oysters. A furious cyclone hit the area on 21 December and threw the men into disarray, separating them and preventing them from finding food. Two men died on Christmas Day, another six soon afterwards.[4] By 6 January 1876, only Baccich and Juric remained alive. An account derived from Baccich described the horrendous events of those 16 days after the cyclone:

> The days passed in great monotony—they had given up all hope of reaching safety. The only ones managing to walk at this stage were Baccich, Jurich [*sic*] and Bucich. In the early morning, before the blistering heat set in, they would search for water and food, but the little they did find was not of great satisfaction. Lovrinovich was falling into a coma and two days after his death Brajcevich and Antoncich passed away. Baccich and Jurich [*sic*] were so weak that they had to support each other to stand. They did their utmost to help Dediol and Bucich. Just over a week had passed since the natives left, when Bucich died. Baccich and Jurich [*sic*] realised the end was nearing—they must get food. Helping one another along they dragged themselves to their dead mate's body and began savagely to tear pieces of flesh from it, swallowing it like animals. 'You poor unfortunates, what are you doing?' gasped poor Dediol, and that was the last he spoke, possibly dying of shock as much as of starvation.[5]

Soon after this, Baccich and Juric were found by the Aborigines, who nursed them back to health. The tribe slowly moved northward to Exmouth Gulf with the hope of making contact with one of the pearling vessels. It was there, on the morning of 18 April 1876, that they met up with Captain Charles Tuckey and the *Jessy*.[6]

When Captain Vinal sailed in to Roebourne in May 1876, his concise message about the *Stefano* angered the community. Local people felt that the Government should provide a vessel specifically to search for shipwreck survivors, and that the lack of such a vessel was costing lives. The schooner *Victoria* was chartered by Government Resident Robert Sholl, who placed the vessel under the command of Pemberton Walcott, with orders to examine the Point Cloates area. Walcott, as an 18-year-old, had survived the wreck of the *Eglinton* a little north of Fremantle in 1852, so his command no doubt elicited exciting memories.

After a protracted voyage, Walcott anchored the *Victoria* inside the reef at Point Cloates at 5 p.m. on 7 June. But when he went ashore, he found more than he had expected, as he reported later:

> After tea being full moon I proceeded on shore in whaleboat accompanied by P.C. Coppin and armed boat's crew—with Tony—native guide. Steering E by N ½ N for a conspicuous hill we landed amid a mass of wrecks. Walked about a mile along the beach and came to the conclusion that several vessels had been lately wrecked—then proceeded on board at 11 p.m.
>
> On the 8th June at 6 a.m. left ship accompanied by same party and Mr Crouch and two natives. Made a minute examination of the different wrecks or parts of wrecks and from differences of wood and size of spars I came to the conclusion that not less than four—probably five vessels of considerable tonnage had been wrecked within seven or eight months. The wreck of *Stefano* was pointed out by natives, and the camp of survivors, where sundry articles belonging to her were found such as boards with written particulars—torn charts, stools etc etc. This wreck, since verified beyond a doubt as the remains of *Stefano* appeared to us the oldest wreck on the beach, as far as amount of damage sustained, and more buried in sand, seaweed etc. But at same time, the paint work appeared brighter and fresher, probably having been newly painted. In her immediate neighborhood on each side within ¼ mile of her were two other vessels, or sides of vessels partially buried in sand and seaweed, and filled up with spars and timber. One of them was Indian built—hard wood—I should judge about from 300 to 500 tons burthen. The other was oregon pine, and apparently American built—about same size. Half a mile further up the beach S. Westerly was the deck of a softwood ship with main hatch, combings etc complete but so choked with spars, masts, yards and other wreckage, as well as to a great extent buried in seaweed, that I found it difficult to identify it as a part of one of the other wrecks. A

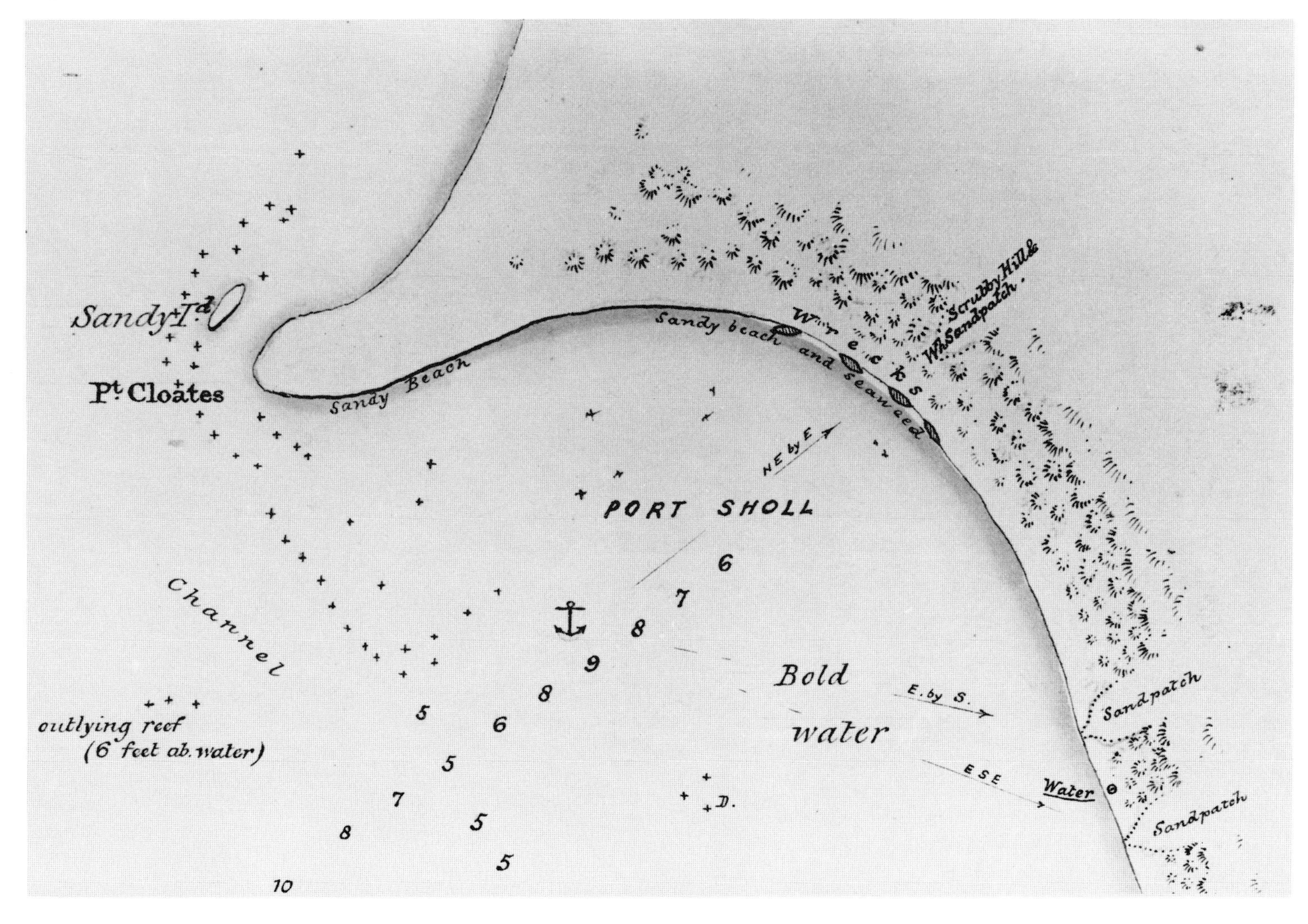

Walcott's chart showing the wrecks on the beach at Point Cloates.

> little further south westerly on the beach, just awash was the side of a very large vessel—apparently lately wrecked hard wood (very like teak) copper fastened and coppered, the copper sheathing being but very slightly torn in a few places and quite bright without any barnacles or other indications of having been any considerable time afloat. I counted 13 planks 9″ wide between her copper and coaming boards. The wood appeared to consist of nearly the whole length of a vessel broadside—but a number of the lower planks had been torn from her timber. There were 13 rows of sheathing left on her and by comparison I am of opinion she must have had about the same number below. 4/5th of the side (being end on) was under water at an angle of about 7° or 8°—the extreme end not being visible. I should judge her to be a vessel of about 1000 tons. I noticed the other side of apparently same vessel afloat about ¼ mile from the beach. No cargo of any kind or boxes etc were observed on the beach. I noticed three built masts of very large size, not less than 3 feet diameter, (a bowsprit I measured was 41 feet long and 2 feet 9 inches dia.), and a main or fore yard, a small portion of one end of which had been broken off measured 71 feet. Hundreds of other spars were strewed along the beach but it was remarkable that all the wreckage had come ashore within a distance of two miles—and literally nothing was seen beyond. The native Tony who had camped with the survivors of the *Stefano* immediately after their wreck seemed very much surprised to see so many other wrecks and declared they were not there at the time he was last there and this has since been verified by statements of the two survivors of the *Stefano* in the *Rosette* who informed me that no wrecks were visible when they left. The native also informed me that about two winters ago a very large steamer had been wrecked down at his country (Cape Cuvier) and all hands lost including a woman.[7]

It is worth reviewing at this point the wrecks that are known or thought to have occurred in the vicinity of Point Cloates prior to Walcott's visit. They are: the 366-ton American China trader *Rapid*, lost just south of Point Cloates in 1811; the Portuguese dispatch vessel *Correo d'Azia*, wrecked in the vicinity in 1816; the 145-ton *Occator*, lost in 1856, probably some distance further north; the 116-ton *Emma*, lost in 1867, a little south at Coral Bay; the 16-ton

Brothers, lost in 1867, possibly in the vicinity; and the small cutter *Bertha*, lost at the Point in 1874. The 450-ton barque *Strathmore* may have struck there in the early 1870s, but floated off. The American galley *Caledonia* struck midway between Point Cloates and Point Maud in 1815, and the galley *Ollices* encountered a reef further north prior to 1818. No other vessels are known to have been lost there between the time the *Stefano* was wrecked and June 1876, when Captain Walcott examined the shoreline.

The beach where Walcott found wreckage acts as a collecting point for flotsam coming over the reef up to 16 kilometres or so to the south of Point Cloates. He would certainly have seen wreckage from the *Stefano* and the *Rapid*, and this would explain his 1,000-ton wreck and the 300–500-ton wreck. The cyclone of December 1875 could have brought ashore, and would have left exposed, material from both of these wrecks, together perhaps with material from the Portuguese wreck. Some element of exaggeration may be seen in Walcott's various accounts. For example, he describes his 1,000-ton wreck as 'not less than 1,500 tons, probably more' in a letter written the previous day, and a week later, he told Sergeant Vincent of the Police at Roebourne that one of the wrecks was of a 2,000-ton vessel.[8] Revelations about large numbers of wrecks suited the popular argument in favour of the Government providing a vessel specifically to assist in searches for shipwrecked sailors. In addition, Walcott needed an impressive report to justify the cost of his protracted voyage. The Government had chartered the *Victoria* at the rate of £4 per day, assuming that the expedition would be of short duration,[9] but Walcott had taken 35 days, costing the Government £140.

Some of the timbers seen on the beach by Walcott would have been removed soon after his visit. Several boats left Roebourne for Point Cloates on hearing Walcott's report, their idea being to salvage the remains. The pioneer pastoralist Julius Brockman came across wreckage on the beach 11 kilometres north of Point Cloates in April 1889. Brockman judged the timbers to be 100 years old, and they may have come from one of the above-mentioned wrecks.[10]

The remains of four of the *Stefano* crew were found a few miles north of

Cape Farquhar in 1880.[11] Some skeletons lie in the hills on Point Cloates itself (where Radovich was buried), but these have not been identified.

In 1981, during a Museum excavation season on the wreck of the *Rapid*, divers saw a trail of coal and wood fragments leading out to sea from the beach where Walcott reported wreckage. It is very likely that the coal is from the *Stefano*. Baccich's account suggests that the *Stefano* struck Black Rock, a detached rock further out to sea than the main reef, but the original wreck site, which should comprise anchors, and perhaps coal and some structure, has not been found.

Baccich told the court of inquiry that the *Stefano* was an 875-ton barque of the Port of Fiume (now Rijeka), the chief port of Croatia in what was then part of the Austro-Hungarian Empire (but is now part of Yugoslavia).[12] An account attributed to Baccich states that the vessel was owned by his uncle and was built at Fiume in about 1873.

Registre Veritas lists this vessel.[13] The 858-ton Austrian barque *Stefano* was built by G. Brazzoduro at the Port of Fiume in 1873, and was owned by N. Baccich and Co. The vessel was built with a single deck and orlop beams of oak, larch and beech, and was bolted together with yellow metal and galvanized iron. She measured 51.9 metres by 10 metres by 6.3 metres. The *Stefano* was surveyed and yellow-metalled in London in June 1875, at which time she was a first-class vessel, with both hull and stores classified as being in first-class condition.

NOTES

1. R. Vincent, Roebourne, 19 May 1876, Report Police Records, Acc. No. 129, Battye Library.
2. Michele Baccich (J. Vincent translator), evidence at Court of Inquiry into the wreck of the *Stefano*, Fremantle, 8 May 1876, C.S.R. 844, fol. 78.
3. Neven Smoje, 'Shipwrecked on the North West Coast: The Ordeal of the Survivors of the *Stefano*, *RWAHSJ* 8: 2 (1978), p. 37. Smoje's article is taken mainly from an account drafted in later years by Baccich.

4. Baccich, evidence at court, fol. 79.
5. Michele Baccich, 'The Wreck of the *Stefano*, translation by Julia Leahy of an article in the Yugoslav magazine *Arena*, Battye Library PR9046.
6. Smoje, *op. cit.*
7. P. Walcott to R. J. Sholl, Cossack, 21 June 1876, C.S.R. 844, fol. 105.
8. Walcott to Sholl, 20 June 1876, C.S.R. 844, fol. 104. See also R. Vincent, Roebourne, 29 June 1876, Report, Police Records, Acc. No. 129, Battye Library.
9. George Crouch and Pemberton Walcott, Memo of Charter, C.S.R. 844, fol. 99.
10. Diary of Julius Brockman, 5 April 1889. We are grateful to Mrs Jean Brockman, of Nedlands, for making us aware of this information.
11. Walter Howard to Inspector of Police at Geraldton, 10 March 1880, Report, Police Records, 29/149, Acc. No. 129, Battye Library.
12. Baccich, evidence at court.
13. *Registre Veritas*, 1875 and 1876.

Gratitude

The 298-ton three-masted schooner *Gratitude* arrived at Fremantle from the Port of East London (South Africa) in ballast. On 24 November 1875, Assistant Pilot W. Simmons took charge of the *Gratitude* to pilot her down to Rockingham, where she was to load a cargo of jarrah sleepers for Port Pirie.[1] The *Gratitude* stranded on the Southern Flats, and 60 tonnes of ballast was discharged before she floated free the next day.[2]

NOTES

1. Register of Arrivals at Fremantle.
2. Fall, *The Sea and the Forest*, p. 71.

Lady Franklin

The 235-ton barque *Lady Franklin* arrived at Fremantle from Adelaide on 21 December 1875, carrying a general cargo.[1] She drifted ashore at Arthur Head on 22 December as a result of the ring of one anchor breaking, and the stock not setting on the other.[2] The vessel was soon refloated, undamaged.

NOTES

1. Register of Arrivals at Fremantle.
2. George Forsyth (Harbour Master) to Col. Sec., 29 December 1875, C.S.R. 813, fol. 254.

The barque *Lady Franklin* at Port Arthur in earlier years.

Wild Wave, Lily of the Lake and Blossom

The pearling fleet in and around Exmouth Gulf was struck by a severe cyclone on 23 December 1875, resulting in the loss of several vessels and fifty-six lives. Pearling masters still had little appreciation of the dangers of operating during the cyclone season, so pearling craft lay fishing in vulnerable positions all along the coast. To the east of Tien Tsin Harbour were the schooners *Good Luck*, *Mary*, *Pearl* and *Venus*, and the cutters *Gypsey*, *Maud*, *Rover*, *Edward James*, *Prince of Wales*, *Firefly*, *Arabian*, *Start*, *May* and *Gift*. The *Mazeppa* was at Flying Foam Passage, the *Nautilus* at the Fortescue River, the *Na Malole* and *Victoria* at Hampton Harbour, the *Adur*, *Mystery* and *Ethel* at Wormbrau Creek, and the *Onward*, *Mira*, *Ione*, *Amy*, *Challenge*, *Waterlily*, *Swan*, *Bessie*, *Argo* and *Marten* at Mary Ann Patch.[1] Other vessels lay further west.

On 22 December at the first squall, the *Morning Star*, *Cygnet*, *Fortescue*, *Twilight* and *Emma* ran in from the Mary Ann Patch to Coolhera Creek. On the morning of the 23rd, the level of the barometer fell from the usual 765 millimetres to 754 millimetres, and by the 24th, had fallen to 734 millimetres.[2] The *Swan*, having parted one chain cable, slipped her other anchor, and at daylight, ran into Coolhera Creek, where she was made fast to the mangroves. At Beadon Creek, the *Subahani* was taking in water and went ashore over the inner reef, high and dry on the beach, but little damaged. (See map on p. 186.)

The coasting schooner *Agnes* had delivered stores to the boats at the Mary Ann Patch on the morning of the 22nd. With the gale increasing, the master got under weigh the next morning and stood out to sea. She was totally dismasted, had her decks swept and was thrown on her beam ends, but did not sink.[3]

The *Blossom* and the *Hope* were anchored in Troubridge Creek at the entrance to Exmouth Gulf. The *Blossom*, a small schooner worked by five Filipino divers, foundered, and all hands were lost. The vessel was later found by a Mr C. Annois as a partly burnt total wreck. The *Industry*, anchored outside Troubridge, dragged about with a short scope of chain as the cyclone veered, but the vessel survived.

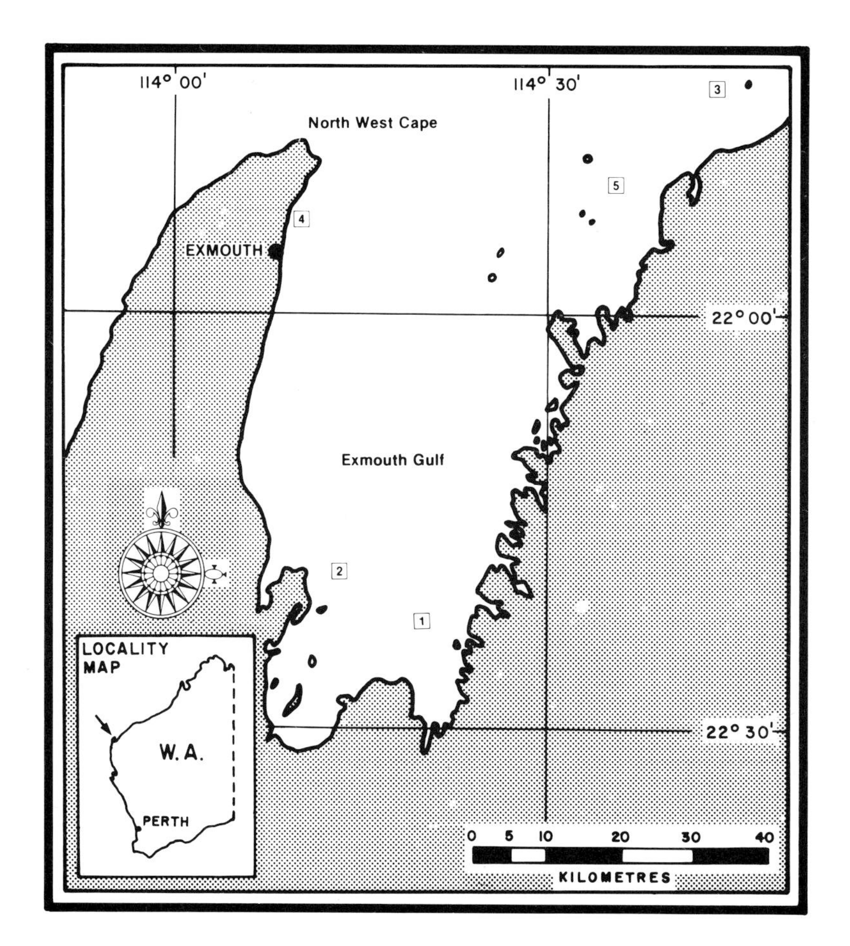

1 *Wild Wave*
2 *Lily of the Lake*
3 *Ariel*
4 *FAIRY QUEEN*
5 *Blossom*

At the southern end of Exmouth Gulf lay the *Dawn*, *Azelia*, *Montiara*, *Lily of the Lake*, *Wild Wave*, *Dolphin*, *Governor Weld*, *Barringarra*, *Charm*, *Ada*, *Aurora*, *Eveline Mary*, *Helena*, *Victorian* and *Albert*. The vessels were on the east side of the Gulf about 3 kilometres off the mainland, and the pearling masters, believing themselves to be well sheltered there, did not run for cover.

The *Albert* was driven ashore in the Bay of Rest, but sustained no damage. The *Lily of the Lake* weathered the first fury of the gale, but during the lull that ensued, her crew unwisely got the vessel under weigh, with the probable intention of running for the Bay of Rest. The next set of squalls came from the westward, dismasting the *Lily of the Lake* and swamping her when only 8 kilometres from the rest of the fleet. The owners, G. Long and W. Woods, together with a European crewman named O'Neil, and seventeen Malay crew and divers were lost when the vessel foundered. Six Malays swam at least 16 kilometres to shore at Mungina, where they were later picked up by the *Barringarra*. It was later found that the *Lily of the Lake* had gone down, head foremost, in 10 metres.

The *Wild Wave*, a well known craft formerly engaged in the coasting trade, had recently been purchased by Aubrey Brown and Charles Gill (part-owners of the ill-fated *Fairy Queen*) for the pearl fishery. The vessel parted her chains in the face of a terrific gale that lashed the sea in the Gulf into one sheet of rolling surf. She went onto a reef between 5 and 6 p.m. on Christmas Eve. Gill and his master, Watson, together with two Europeans named Shadwell and Cameron and twenty-eight Malays, were drowned.[4] Gill clung to the masthead for several hours before succumbing. Thirteen Aborigines and one European named Kennington survived by clinging to the rigging overnight and swimming ashore the next day. Kennington was found lying unconscious on the beach, and he remained delirious for at least 3 days.[5]

The 28-ton *Wild Wave* (Official Number 40482) was built in 1858 at Fremantle as a wood-framed ketch with a square stern. In 1864, the vessel was converted to schooner rig.[6] Her dimensions were 15.4 metres by 4.6 metres by 2 metres. She had been owned by John and Walter Bateman as late as July 1875.

The 26-ton schooner *Lily of the Lake* (Official Number 61108) was built in 1873 at Fremantle by James Storey for storekeeper John Lewis, who later sold her to William Owston.[7] The jarrah vessel had two masts, a round stern and the dimensions 17.7 metres by 4.2 metres by 2.1 metres. The *Lily of the Lake* was used initially in the coasting trade. In February 1874, she was for a time feared lost off Geraldton, and the local policeman wrote:

> Captain Mitchell informed me she had large quantity of beer and spirits on board and that the passengers were rather a rough lot.[8]

In mid-1875, the *Lily of the Lake* was taken to Coepang to recruit men for a pearling cruise. When she was boarded by a quarantine officer at Roebourne, he reported that, besides the master and mate, the vessel carried twenty-one Malay divers and crew, one cook, and one woman, whose function on board was not given.[9]

Virtually nothing is known of the *Blossom*. The vessel was variously described as a cutter and a small schooner.

NOTES

1. *Herald*, 29 January 1876.
2. *Western Australian Times*, 14 January 1876.
3. Board of Trade Wreck Register, 1875.
4. *Herald*, 29 January 1876. The report of the Roebourne police (22 January 1876) implied that the men lost were Shark Bay Aborigines.
5. *Inquirer*, 12 January 1876.
6. Register of British Ships, Fremantle.
7. Register of British Ships, Fremantle.
8. Joseph Armstrong, Geraldton, 4 February 1874, Police Records, Acc. No. 129, Battye Library.
9. Government Resident, Roebourne, to Col. Sec., 7 December 1875, C.S.R. 809, fol. 183–187.

Star

The cutter *Star* left Shark Bay on 5 April 1876, bound for Geraldton, where Charles Brennan, the master, was intending to buy provisions.[1] Brennan, who was well acquainted with the coast, had two crewmen and a passenger on board with him. The passenger, John Hill, was making his way to Fremantle with a valuable lot of pearls. The vessel carried provisions for 7 to 10 days.

By mid-May, the vessel's non-arrival was causing concern. The Resident Magistrate at Geraldton reflected that the wind had blown heavily from the east and south-east after the *Star* left Shark Bay.[2] Rumour circulated that smoke had been seen in the direction of the Abrolhos, but a search of the islands by the *Waterlily* revealed nothing.[3] The *Star* was a light boat, and it was feared that she had capsized. (See map on p. 53.)

The *Star* was a 5-ton cutter that had probably been employed in pearling at Shark Bay. It is likely that she was not registered. A 16-ton cutter of the same name (Official Number 31500) was registered at Adelaide in the 1870s.[4] That vessel was 9.4 metres by 3.5 metres by 1.8 metres.

NOTES

1. Resident Magistrate, Geraldton, to Col. Sec., 13 May 1876, C.S.R. 844, fol. 89.
2. Resident Magistrate to Col. Sec., 13 May 1876.
3. Telegram from Resident Magistrate, Geraldton, to Col. Sec., 15 May 1876, C.S.R. 844, fol. 89.
4. Register of British Ships, Adelaide.

Speculator

More deaths occurred in the pearling industry in May 1876. On the 12th, the two small vessels *Speculator* and *Aurora* started to beat out from a small island between Legendre and Gidley Islands. On board the *Speculator* were Henry McCaffray, James McGee, Joseph Rogers and John Spencer. The *Speculator* was not making headway against the tide, so Spencer took the dinghy ahead, sculling past an adjacent point of land. At about 3 p.m., the men remaining on board let out the reefs from their sails and were going about when an unexpected squall of wind capsized the vessel.[1] (See map on p. 118.)

The *Speculator* was built of light wood and did not sink, but the stores all shifted to one side, and as the crew had nothing with which to cut the mast and sails away, they were unable to right her. At the time of the accident the *Speculator* was only about 200 metres from the nearest point of land, but now the tide was taking the wreck into the middle of the stream. McGee left the wreck to swim ashore, but drowned. McCaffray started out towards the shore, supporting Rogers, who could not swim. But the tide was too strong, so McCaffray eventually left Rogers standing on the wreck with water up to his chest, and swam to Gidley Island. After crossing two or three channels and walking about 20 kilometres, he found the *Aurora*. By then, it was quite dark, and a strong wind had brought on heavy seas.

With the morning high tide, the *Aurora* started back towards the place where the *Speculator* had capsized, but the wind was dead ahead and they had not reached the site by sundown. The following day, the winds were too high, so the search was abandoned.

McCaffray described the *Speculator* as an open boat that would carry about 2.5 tonnes, and referred to its one mast.[2] The Government Resident at Roebourne referred to it as a 2-ton boat.[3]

There was apparently a larger vessel of the same name in the Colony—a 15-ton schooner and cutter (Official Number 40485). This larger vessel was the rebuilt hull of the paddle-steamer of the same name launched in 1854, and it may be expected that that vessel would have been lightly built for river work.[4] Its dimensions were 13.4 metres by 3.7 metres by 1.5 metres. *The Mercantile Navy List of Australia and New Zealand* shows that the cutter was built at Fremantle in 1859 and owned by R. F. Phils in 1871.[5]

NOTES

1. Henry McCaffray, evidence given to Frederick Pearse, Justice of the Peace, 17 May 1876, C.S.R. 844, fol. 101.
2. McCaffray, evidence.
3. Government Resident, Roebourne, to Col. Sec., 19 May 1876, C.S.R. 844, fol. 103.
4. *Inquirer*, 18 October 1854. See also *Inquirer*, 9 May 1883.
5. F.C. Jarrett, *The Mercantile Navy List of Australia and New Zealand* (Sydney, 1871).

A steamboat on the Swan River in 1859.

Gem

Some time between 6.30 and 7.00 a.m. on 18 May 1876, John Hore, the Assistant Lightkeeper on Rottnest Island, saw a vessel about 3 kilometres east of the island. Although it was still fairly dark and hazy, he recognized her as the cutter *Gem*, a coasting vessel and therefore not in need of a pilot. Hore was then occupied for about three quarters of an hour extinguishing the light.[1] When he turned his attention back to the water, the *Gem* had disappeared, and he concluded that the vessel had sailed into Fremantle. (See map on p. 242.)

The Lightkeeper at Arthur Head (Fremantle) also noticed the vessel and looked away. But when he returned his attention to the vessel, he saw only a mast sticking up in the water. He relayed this information to Harbour Master George Forsyth, who hastened away towards Rottnest with his boat. Forsyth was some time in making the passage, because of a strong northerly wind and a heavy sea.[2] He found the *Gem* lying on the bottom, resting on her starboard bilge, her mainsail set but the topmast carried away at the cap. The dinghy was floating with its stern just above the water, the bow held down by the painter.

Forsyth went on to Rottnest, and not finding the crew there, he returned to the wreck, where he retrieved a coat from the mast-head. He conducted a search of Carnac Island, Mewstone Rock and Stragglers Rocks, without finding any of the crew.

The next day, the cutter *Two Sons* was sent out with several divers to examine the hull and recover any trapped bodies. The divers got into the hold and examined the captain's cabin, but the entrance to the main cabin was small and cluttered with rigging, so they were afraid to go in. The divers were also frightened by 'a great many large bloated looking sharks'.[3]

None of the ten men on board the *Gem* were found, and this fact has led to various theories being raised as to the cause of the wreck. One is that the *Gem* had sprung a small leak early in the voyage, and that this caused the cargo of wheat to swell, leading to a sudden bursting of the ship's hull and consequent immediate sinking.[4] Another theory is that the *Gem* struck the nearby Kingston Reef and foundered as a result of being holed.[5] It has been suggested that the Harbour Master was involved in some sort of cover-up, because of dis-

The Fremantle Harbour Master's boat heading out to sea.

crepancies in the stated depth and position of the wreck. However, the Harbour Master had made his estimates from a small boat in heavy seas, and besides, the rigged hull of the *Gem* could well have bounced along the seabed a good distance before coming to her final resting place. In 1878, the *Inquirer* discounted reports that the crew of the *Gem* had been discovered living on an island on the North West coast.[6]

The *Gem* had left Port Irwin at 6 a.m. on 17 May with a cargo of 500 bags of wheat. On board were Captain Wilcock; the mate, W. Smith; the cook Edward Kying; seamen Garrod, Kingslake and Haggensen; and passengers Joseph Johnston, William Faulkner, Henry Chattfield and George Jones.[7]

The wreck of the *Gem* broke up quickly with the swelling of the wheat, and the cabin top, bulwarks, pieces of the deck and other items were found on the beach at Fremantle 3 days after the sinking. In these circumstances, it is unlikely that any bodies would have remained in the hull for long.

It seems that the location of the wreck of the Gem has been known for many

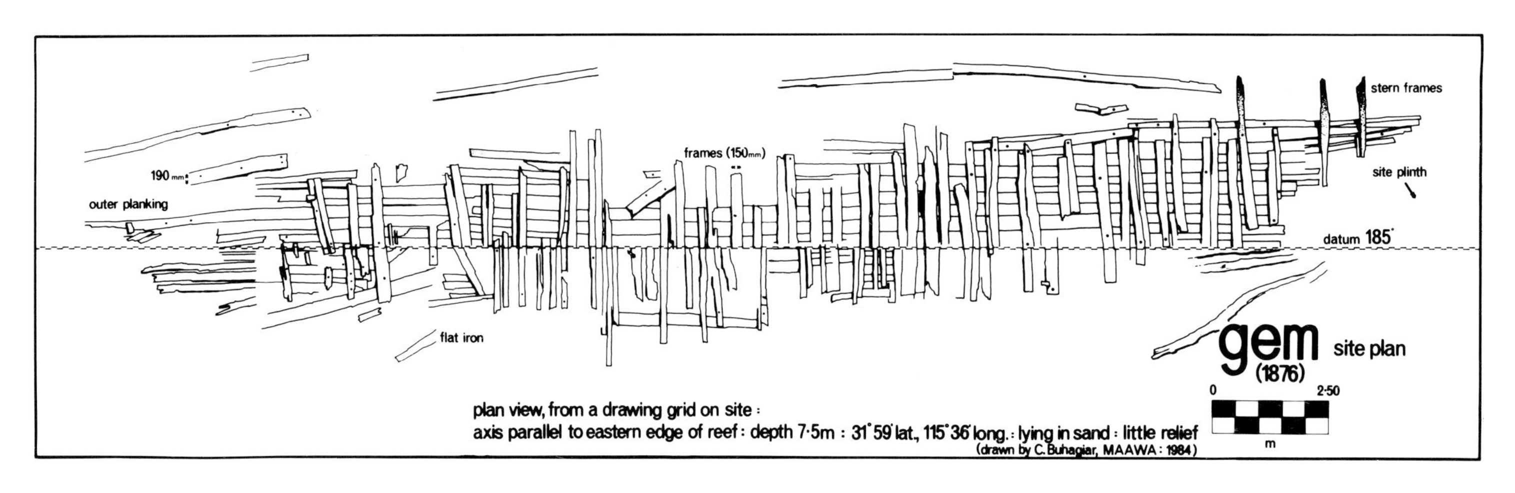

Site plan of the *Gem*, by the Maritime Archaeology Association of Western Australia.

years. The site lies on sand bottom in about 10 metres of water, approximately a kilometre north-east of Phillip Rock, in latitude 31° 59.6′ south, longitude 115° 33.5′ east, and consists of a keel, with ribs and substantial planking. In 1984, the Maritime Archaeology Association conducted a survey of the wreckage. The site is protected under the Historic Shipwrecks Act.

The 52-ton cutter *Gem* (Official Number 31520) was built at Cowes on the Isle of Wight in 1835, so it was 41 years old when it sank at Rottnest.[8] The vessel was carvel built, with one deck, no figurehead, a square stern and the dimensions 20.1 metres by 4.8 metres by 2.6 metres. It was owned in Adelaide and Melbourne before being sold to Fremantle for the coastal trade.

NOTES

1. John Hore (Assistant Lightkeeper) to Col.Sec., 26 May 1876, C.S.R. 830, fol. 126.
2. George Forsyth (Harbour Master) to Col. Sec., 18 May 1876, C.S.R. 844, fol. 95.
3. Sergeant Joseph Campbell, Report, 19 May 1876, Police Records, Acc. No. 129, Battye Library.
4. W. Somerville, *Rottnest Island, Its History and Legends* (Rottnest Island Board, 1966), p. 113.
5. This theory is held by Maritime Archaeology Association members who surveyed the site in 1984.
6. *Western Australian Times*, 26 May 1876.
7. *Inquirer*, 21 August 1878.
8. Register of British Ships, Melbourne.

Bungaree

The schooner *Bungaree* left Batavia for Fremantle on 21 May 1876 under the command of John Cornford, with a European mate and six Malay crewmen. The vessel was laden with about 32 tonnes of sugar and sundries.[1]

Favourable weather prevailed until 10 June, when the barometer began to fall. Captain Cornford fixed his position by morning sight that day, but failed to get a meridian altitude for latitude, and was unable to get any sights after that date because of the weather. He continued on by dead reckoning, steering east on the 10th, and east by north on the 11th. On the 12th, Captain Cornford reckoned that he was 315 kilometres west of Rottnest Island, so he continued east by north.

The mate gave warning 3 minutes before the vessel struck breakers on the starboard bow, at 4.30 a.m. on 13 June. Captain Cornford flung down on the helm and the vessel swung round to north-east, but when she struck, the bottom was stove in and she filled at once. Two hands went to the pumps, while the remainder of the crew readied the boats for launching.

Not knowing where they were, the crew remained on board until 8.30 a.m., But as the light improved, they saw Long Point about 3 kilometres away, and realized that they had been wrecked on the Sisters Reef. Twice the boat was swamped by the breakers and had to be hauled back in, but on the third attempt, the crew got clear of the wreck and landed in safety on the Point. The sextant and a tin box containing the ship's papers were saved.

Captain Cornford walked to Rockingham, where he obtained a trap for the journey to Fremantle. The next day, he returned to the site of the wreck, but the vessel had completely broken up, and the beach was strewn with debris.[2] A boat belonging to Messrs Bateman, the owners, was loaded with coconuts and pomelos from the shore, and headed back towards Fremantle. Bad weather forced the three crew to shelter behind Penguin Island, and during the night, the boat dragged ashore and was stove in.

At an inquiry into the cause of the wreck, Captain Cornford was charged with neglect of duty in omitting to take soundings, and with want of caution and neglect in standing in for the land in thick weather. In evidence, the mate, Robert Kirk, who had been on the *Bungaree* for 11 months, stated:

> there was a deep sea lead on board, could not tell the length of it, did not ever use it while I have been on board.[3]

The court suspended Captain Cornford's certificate for 3 months.

The wreck of the *Bungaree* has never been found. It was a small vessel, and it broke up quickly, but the records give the location reasonably well. Captain Cornford said that the vessel was about 3 kilometres from Long Point (the *Herald* said about 2 kilometres), and the *Western Australian Times* placed the wreck 0.9 kilometres south of the Sisters Rock.[4] (See map on p. 242.)

The topsail schooner *Bungaree* (Official Number 38796) was built at Jervis Bay, New South Wales, in 1866. The carvel-built vessel had one deck, two masts, a round stern and a male bust figurehead.[5] Its dimensions were 25.8 metres by 5.9 metres by 2.5 metres. In 1868, Walter Bateman, of Fremantle, bought the *Bungaree* from William Young, of Melbourne, for £1,575. The vessel was not insured.

NOTES

1. John Cornford, evidence at inquiry into *Bungaree*, 21 June 1876, C.S.R. 844, fol. 121.
2. *Herald*, 17 June 1876.
3. Robert Kirk, evidence at inquiry.
4. *Herald*, 17 June 1876. See also *Western Australian Times*, 16 June 1876.
5. Register of British Ships, Melbourne.

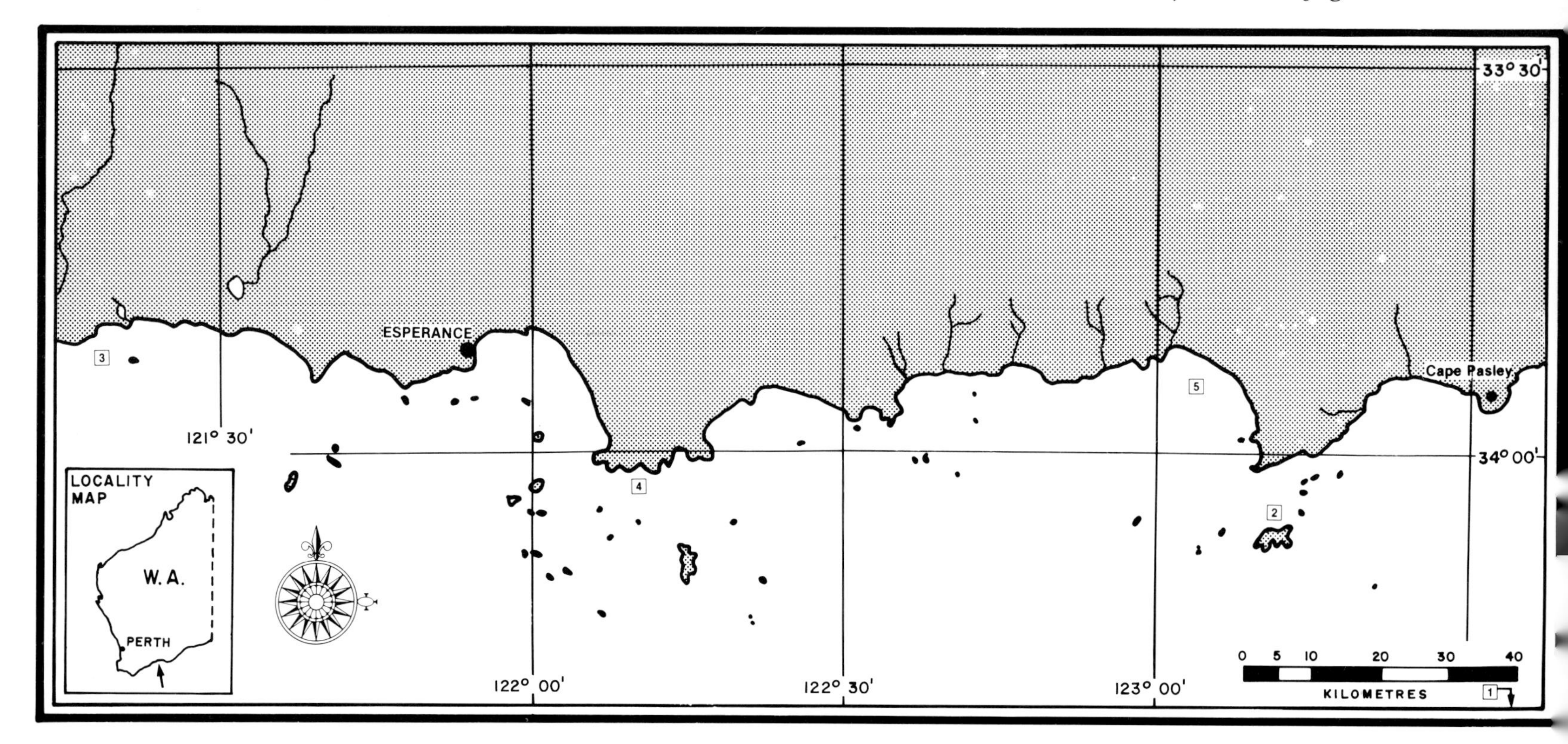

1 SS *Start* on Pollock Reef
2 *Belinda*
3 *Mary Ann*
4 *Mountaineer*
5 *BATOE BASSI*

Mary Ann

The 104-ton topsail schooner *Mary Ann* was built at North West Bay in Tasmania in 1849, and was owned in Hobart during the early 1850s. The mariner and grazier Captain James Dempster bought the schooner in 1868 and had her fitted out in Melbourne for the North West (Western Australia) pearl fishery.[1] However, when the vessel arrived at Fremantle, he also contemplated using her for collecting bêche-de-mer, and he later traded with her to Mauritius.

In April 1869, the *Mary Ann* struck a rock at Rocky Passage, in Flying Foam Passage, and filled with water.[2] But although holed, the hull was buoyant and drifted onto sand bottom, where salvage was feasible. George Howlett bought her to trade at ports between Fremantle and Singapore. In 1871, for example, she carried mother-of-pearl shell, tortoise-shell and sandalwood to Singapore, and sugar to Fremantle.[3]

Times were hard in the Colony in the years immediately after the cessation of transportation, and Howlett had to struggle to survive as a shipowner. In July 1871, the *Mary Ann* was driven ashore at Champion Bay and damaged, and he had to throw most of his cargo overboard. Yet he could write stoically to his agent:

> We cant get a fair wind, it is still blowing from the N.W., the glass falling, but I am not afraid but that I will do well with the old craft yet if I can only get out of this place.[4]

Howlett bought the brig *Wild Wave* in May 1872, but a month later, he was served with a bottomry bond on the *Mary Ann*.[5] A bottomry bond is a pledge of a ship as a repayment of a debt. Ever on the lookout for new ventures, he sent her on a whaling voyage to Rosemary Island in the hope of paying off the debt to his agent. In August, he wrote:

> With regard to the *Mary Ann* as soon as she has finished whaling if Bateman is willing I would like her to remain as she is so as to be ready for next season, and put about 30,000 ft of timber such as different sized boards, quartering battens, and about 20 tons of good flour and anything

> else you might think of and make a store ship of her and keep her on the pearl banks. We would then get shells as ready money and I would not have as many bad debts.[6]

Presumably that scheme did not work, because the vessel was sold in 1874 for £405.[7]

In 1875, the *Mary Ann*, now owned by Messrs Pearse and Owston, began conveying telegraph poles to the head of the Great Australian Bight for use in the construction of the Eucla telegraph line. The vessel was chartered to William Miles and commanded by John Christie. In May 1876, the *Mary Ann* was engaged to take some officials and plant to Eucla. That task completed, Captain Christie left Israelite Bay on 12 July for Middle Island, but progress being slow, he decided to anchor the next evening under the lee of Bellinger Island.[8]

The *Mary Ann*, from a crayon drawing, *Muresk College Magazine*, 1928.

The wind direction changed overnight, putting the *Mary Ann* on a lee shore. In getting under weigh on the 13th, Captain Christie hove the anchor cable short and set his topsails, but the *Mary Ann* would not cant, and began to drag her anchor closer to the shore. The sails were furled and extra cable let out to stop the vessel, but attempts to warp her further out to sea were unsuccessful. In the evening, a fresh north-easterly wind combined with a heavy swell to push the vessel ashore, dragging her 900-kilogram anchor behind her. The six crew and six passengers got on shore safely with their personal effects, but the 27-year-old *Mary Ann* became a total wreck.[9]

The loss of the *Mary Ann* did not affect the schedule for completion of the line. In 1877, Perth was connected, for the first time, by telegraph with London via Albany, Esperance, Israelite Bay, Adelaide and Darwin.

The *Mary Ann* (Official Number 31932) was a wooden sailing vessel of carvel build, with standing bowsprit, square stern, scroll head and no galleries. She had one deck, two masts, and the dimensions 24.7 metres by 5.9 metres by 2.8 metres.[10]

NOTES

1. *Inquirer*, 13 May 1868.
2. John McKenzie (master), statement regarding wreck of *Mary Ann* , Roebourne, 18 May 1869, C.S.R. 647, fol. 45.
3. See The Southern Insurance Co. Ltd, Policy 257, and Bill of Lading, *Mary Ann*, 8 November 1871, Habgood Papers, 813A, Battye Library.
4. George Howlett to John Absolon, 24 July 1871, Habgood Papers.
5. Bottomry Bond, *Mary Ann*, 8 June 1872, Habgood Papers.
6. Howlett to Absolon, 19 August 1872, Habgood Papers.
7. Lionel Samson to G. Crouch, 19 June 1874, Samson Papers, Acc. 2169A, Battye Library.
8. William Miles, evidence, preliminary Court of Inquiry into the stranding of the Schooner *Mary Ann*, 29 August 1876, C.S.R. 844, fol. 155.
9. John Christie, evidence.
10. Hobart Customs Register, CUS 38 (Tasmanian State Archives).

Hero of the Nile

The barque *Hero of the Nile* was in sand ballast when she set off from Melbourne on what was to be her last voyage in 1876. Under the command of Captain N. H. Dugdall, the vessel was bound for the Lacepedes to pick up a cargo of guano.[1]

At 5 p.m. on 19 October, Captain Dugdall estimated his position to be about 55 kilometres off Cape Bouvard. With a south-westerly behind him and full plain sails set, Dugdall steered north-east-half-east, travelling at about 11 kilometres per hour. At 8 p.m., he altered the course to north-east, intending to steer for the Rottnest lighthouse. Then at 1 a.m. on the 20th, the man on the lookout called 'land under the lee', so the course was changed to north-west.[2]

Expecting that his new course would take him clear of the land, Dugdall did not bother to take soundings. He had with him a general chart, which did not show the dangers of the area. Currents had taken the *Hero of the Nile* inside the Murray Reef, to a very dangerous position.

The *Hero of the Nile* struck Long Point at about 2 p.m. Captain Dugdall backed the sails, but finding this to be of no use, he furled them. The vessel bumped very heavily all night, and was making a great deal of water, and the pumps were soon choked with sand ballast. After daylight, an anchor and warps were run out, but it was of little use, because the vessel was full of water. Captain Dugdall, his thirteen crew and two passengers, one of whom was his wife, all got ashore in the ship's boat. (See map on p. 242.)

Captain Johnson, of the cutter *Eveline Mary*, was making his way up to Fremantle from Bunbury when he saw the stranded barque at about 8 a.m., and he provided Dugdall and his wife with transport to Fremantle.[3] When the Fremantle Harbour Master visited the wreck the next day, it was lying some 275 metres from Long Point, canted right over on its port side and held down by the sand ballast.

The *Hero of the Nile* was condemned by a board of survey and sold as a wreck at public auction. The hull was knocked down to Messrs Higham and Sons for £100, and the gear and furniture were sold in small lots, bringing the total to about £500.[4] The court of inquiry found that no blame could be attached to Captain Dugdall, because the available charts of the area were not good enough for close navigation.

The wreck was shown on an 1879 chart of Warnbro Sound, and its position was known by local fisherman Mr Broz, who led Harold Roberts and other members of the Underwater Explorers Club to the site in about 1960.

When Museum staff inspected the site in 1974, the timbers were lying on sand and weed bottom at a depth of 2.4 metres, some 300 metres north-west of Long Point at latitude 32° 22′ south, longitude 115° 42′ east. The site is protected under the Historic Shipwrecks Act.

The 356-ton barque *Hero of the Nile* (Official Number 4668) was built of wood at West Cowes in 1852, and owned by Thomson and Westmoreland of London.[5] The vessel was 38.4 metres long by 7.6 metres wide by 5.2 metres deep.

NOTES

1. Captain N.H. Dugdall, evidence, Preliminary Inquiry into the loss of the barque *Hero of the Nile*, 30 October 1876, C.S.R. 844, fol. 166.
2. Captain Dugdall, evidence.
3. H. Mills, Report, 20 October 1876, Police Records, Acc. No. 129, Battye Library.
4. *Herald*, 4 November 1876.
5. *Lloyds Shipping Register*, 1876.

Annie M. Young

The brig *Annie M. Young* arrived at Bunbury on 6 October 1876 from the Vasse, where she had taken on fifty loads of jarrah for New Zealand. After many delays caused by strong westerly winds, she took on 170 loads of long piles and, on 31 October, was ready to go to sea.[1] Captain Samuel Tiddy weighed anchor on 1 November, but the wind fell, so after moving out only a short distance, Captain Tiddy anchored again. Then he unwisely went ashore, leaving his vessel in a more exposed position.

The following day, the wind blew strongly from the west, at night increasing to a gale from the north-west. Soon after midnight, the *Annie M. Young*'s

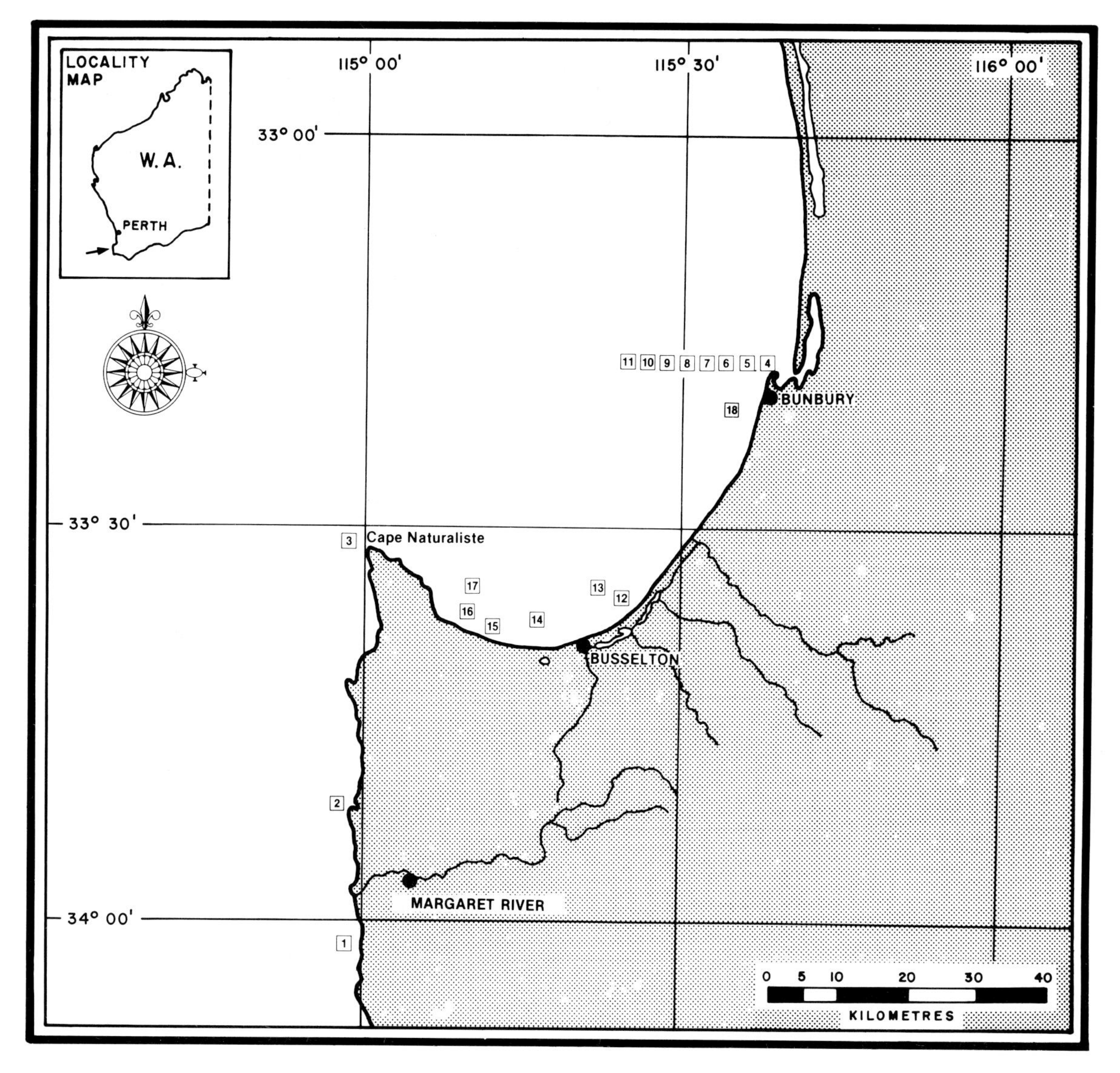

1 SS *GEORGETTE*
2 *Thetis*
3 Unnamed Boat 1875
4 *Citizen of London*
5 *Perseverant*
6 *Annie M. Young*
7 *Midas*
8 *Samuel Wright*
9 *North American*
10 *North American*
11 *Elizabeth*
12 Boat 1801
13 *Mary*
14 *Governor Endicott*
15 *Halcyon*
16 *Ella Gladstone*
17 *Geffrard*
18 *Eliza*

chain cable parted. Two more anchors were let go, but both cables parted. The crew sent up a distress rocket and burnt blue lights, but no one could assist. She drove ashore on the north beach, stern on, and soon bedded into the sand with less than a metre of water at the bow, and the hold full of water to the level outside. The cutter *May* was blown ashore nearby, but sustained no serious damage.

Captain Tiddy hurried across the estuary to the north beach and found the *Annie M. Young* ashore 2.4 kilometres from the mouth of the estuary. From the shore, he ordered the crew to set the lower fore topsail and the fore top-mast staysail, to force the vessel further up on the beach.[2]

On 4 November, surveyors recommended that the timber piles, valued at over £1,000, be discharged, and, later in the month, it became apparent that the vessel was beyond recovery. The cargo was insured, but the vessel was not. An inquiry into the stranding resulted in the captain and crew being acquitted.[3]

The 303-ton brig *Annie M. Young* (Official Number 48084) was built at Yarmouth, Nova Scotia, in 1863 by and for Messrs Young and Baker.[4] It was later sold to W. McCormick, of Dublin. The vessel's dimensions were 33.2 metres long by 8.1 metres wide by 5.3 metres deep. Lloyds Survey Register shows that the vessel's frames were of iron, the rudder of oak, the keel of birch and maple, the bottom planking of birch and spruce, the floors of birch, beech and tamarack, and the treenails of beech and tamarack.[5] The *Annie M. Young* was equipped with two Trotman's design bower anchors, two stream anchors, and two Trotman's kedges.

NOTES

1. W.P. Clifton (Resident Magistrate, Bunbury) to Col. Sec., 4 November 1876, C.S.R. 844, fol. 190.
2. Samuel Tiddy, evidence, C.S.R. 844, fol. 200.
3. *Inquirer*, 29 November 1876.
4. Lloyds Survey Register, National Maritime Museum, England.
5. Lloyds Survey Register.

Kitty Coburn

The brigantine *Kitty Coburn* first arrived at Augusta, in Flinders Bay, on 12 November 1875. Four days later, she was in Albany taking on passengers before returning to Augusta for a load of timber intended for Adelaide. But at 6 p.m. on 3 December, a strong gale from the south-east parted both of the *Kitty Coburn*'s cables and she was driven ashore about 9 kilometres north of Barrack Point, which lies directly seaward of the Augusta town site.[1]

At the inquiry, Captain W. Steele claimed that Flinders Bay was very dangerous in a south-easterly gale. The *Kitty Coburn*'s hawse-pipes had been ripped out before the chain cables broke.[2] It was thought at the time that the vessel would become a total wreck. The Captain did not know whether the ship, which contained 130 loads of valuable jarrah, was insured.

The wreck was sold by auction in January 1876 for £270, to a group of Fremantle residents. One of them, William Owston, made a careful examination of the hull, and decided that it could be refloated and taken to Fremantle for complete repairs and refitting. A team of workmen, under the supervision of W. Brown and equipped with heaving-down gear, was sent to Augusta on the schooner *Brothers*. The *Kitty Coburn* was refloated, and when Brown saw how effectively the pumps were lowering the water in the hold, he began to feel that it might not be necessary to careen the vessel before taking her to Fremantle.[3]

On 1 November 1876, Brown took advantage of an easterly breeze, and the *Kitty Coburn* set out, in company with the *Brothers*, for Fremantle. But the glass was falling as the anchor was weighed, and the following day, a heavy gale from the north-west was experienced off Cape Leeuwin. The vessel began to leak very badly, and at 4 a.m., the crew abandoned her. The *Brothers* made a course for Fremantle, but was driven so far to leeward by the heavy weather that she ran for Albany. The cost of the loss of the *Kitty Coburn* was estimated at £1,600.[4] (See map on p. 20.)

The 405-ton two-decked brigantine *Kitty Coburn* (Official Number 64797) was built of oak, pine, hackmatack and pitch pine at Damariscotta, in the United States, in 1865.[5] Her dimensions were 38.1 metres by 9 metres by 5.1 metres.[6] The vessel was owned by John Smith, of Melbourne.

NOTES

1. Captain W. Steele to I. Harris, Augusta, 3 December 1875, C.S.R. 795, fol. 157.
2. Captain Steele, evidence at Inquiry into stranding of *Kitty Coburn*, Flinders Bay, 11 December 1875, C.S.R. 795, fol. 164–167.
3. *Herald*, 11 November 1876.
4. *Ibid*.
5. *Mercantile Navy List, 1876*. Although described in the *Mercantile Navy List* as a brigantine, the Board of Trade Wreck Register refers to her variously as a schooner and as a barque.
6. *Registre Veritas*, 1870.

Georgette

The development of coastal steam services in Western Australia was first given serious consideration after the cessation of convict transportation in 1868, when commercial interests in Perth and Fremantle felt painfully aware of their insecure future. For the isolated Colony to develop, one of the most important prerequisites was an efficient system of seaborne communications and transport. The ocean mail steamers called at Albany to re-coal *en route* from Europe to the Eastern Colonies, but that was of limited use to the people of Perth and Fremantle, because Cape Leeuwin was such an obstacle to speedy regular communication from Albany to Perth.

Real progress did not take place until May 1872. In that month, the *Xantho* arrived at Fremantle, and in a separate development, Messrs Connor and McKay submitted a proposal to the Government to carry mails on a contract basis. Not surprisingly, the *Xantho* had a short life on the Western Australian coast. Connor and McKay obtained a more suitable vessel, the *Georgette*, and succeeded in establishing under contract a regular steam service between Albany and Champion Bay.

Connor bought the near new 211-ton iron screw-steamer *Georgette* in England for £14,000.[1] The vessel first arrived at Fremantle in September 1873, and began her coastal trading service with a voyage to Champion Bay. In October 1873, she was stranded on the infamous Murray Reef while *en route* from Albany to Fremantle, and was sent to Adelaide for an overhaul. Arriving back in Fremantle in March 1874, she resumed the service, with no major interruptions until December 1876, a period of about 3 years. During that time, she made several trips to Adelaide for overhauls, but attempted to fit these into the service by carrying mails, passengers and cargo there and back. The contract with the Government expired in September 1876.

The *Georgette* established regular trade connections for the first time between Albany, Champion Bay and intermediate ports, and encouraged passengers to travel a good deal more frequently. The vessel's performance in regard to mails was much maligned by the Government, but the Colony found, after the *Georgette* was wrecked, that few other companies were prepared to accept the challenge; the strictly scheduled carriage of mails round the Leeuwin was a formidable task for any Australian coastal shipping company at that time.

The *Georgette* left Fremantle, on what was to be her last voyage, on 29 November 1876. She carried fifty passengers and a cargo of jarrah and sundries, and was bound for Adelaide via Bunbury, Busselton and Albany.[2] The vessel cleared from Busselton on the afternoon of the 30th and all went well until midnight, when the *Georgette* was about mid-way between Cape Naturaliste and Cape Hamelin. A leak developed and neither the ship's pumps nor the donkey steam pump would work. (See map on p. 204.)

At 4 a.m., the water was rising fast, so Captain John Godfrey put all the passengers and crew to work bailing with buckets, while he steered for the coast. At 6 p.m., the Chief Engineer reported that the fires in the stokehole had been extinguished. The drifting vessel was still some kilometres from shore. Captain Godfrey ordered all the boats to be swung out from the davits ready to receive passengers and crew. The first boat was fully loaded with twenty people when a disaster occurred, as Mrs Annie Simpson, one of the occupants, later recalled:

> Just then another big wave struck the lifeboat and threw it against the ship's side. It broke clean in half from end to end, and I shall never forget the awful screams that went up. All in a few seconds we were struggling in

> the water, and it was then that most of the lives were lost. I floated out on my back, saw the broken boat turn over and float away, and saw the *Georgette* for the last time.[3]

Mrs Simpson was rescued by the second lifeboat and landed at Quindalup, but twelve people drowned.

Meanwhile, the *Georgette* drifted into the surf at Calgardup Bay, and Captain Godfrey immediately commenced landing the remaining passengers with the third lifeboat.

Newspapers created a legend around the circumstances of the landing. Under the title 'A Western Australian Grace Darling,' journalist H. C. Barrett wrote:

> . . . The boat swamped, they were all in the water, and in the greatest danger, when, on the top of the steep cliff appeared a young lady on horseback. Those who were present have told me that they did not think a horse could come down that cliff, but down that dangerous place this young lady rode at speed; there were lives to be saved, and, with the same fearless and chivalrous bravery that urged Grace Darling to peril her life for fellow creations, and gave her a name in all English history hereafter, Grace Bussell rode down that cliff, urged her horse into the boiling surf, and out beyond the second line of roaring breakers, till she reached the boat where the women and children were in such peril. Her horse stumbled over the rope and she was nearly lost, but managed to get alongside the boat, and then with as many women and children clinging to her and the horse as possible, she made for the shore and landed them. A man was left on the boat, and he could not get on shore till Miss Bussell sent her black servant on horseback to aid him. So furious was the surf that it took 4 hours to land 50 people, and every boat engaged was capsised.[4]

Aboriginal stockman Sam Isaacs had first brought the alarm to the Bussells' homestead, and he played a significant role in the rescue. Grace Darling was an English lighthouse keeper's daughter who, in 1838, rescued nine survivors of the wreck of the steamer *Forfarshire*.

When news of the disaster broke in Perth, the community was outraged, and the court of inquiry commenced amid rumours that the *Georgette* had

A romanticized version of the rescue of the SS *Georgette*'s crew, *Illustrated Sydney News*, 3 February 1877.

been deliberately scuttled. Captain Godfrey was charged with the following counts:

1. That without due regard for the safety of the ship he took in a large quantity of jarrah timber hastily, violently and incautiously, thereby injuring the vessel and causing her to leak, resulting in the loss of the vessel and lives of passengers.
2. That he proceeded to sea with an insufficient number of boats, these not being seaworthy.
3. That he placed passengers in a leaky boat, the upsetting of which caused the loss of several lives.
4. That his Chief Officer did not have a certificate of competency or service.
5. That the ship's pumps were damaged and not in good working order.[5]

Captain Godfrey was found not guilty of all these charges, but the court, probably mindful of public opinion, retained his certificate on the grounds that he should have fully ascertained the condition of the *Georgette* early in the evening of the disaster.[6]

The hull of the *Georgette* was sold for £40.[7]

The site of the *Georgette* wreck is today well known, and it is marked on shore by a plaque and cairn at latitude 34° 02′ south, longitude 140° 59′ east. The wreck lies in a depth of 5 metres in the surf line, bows on to popular Redgate (Calgardup Beach), about 90 metres offshore on an axis of 100°. When inspected by Western Australian Museum staff in 1980, the hull outline was barely visible.[8] From the stern-post, there appeared a short propeller shaft terminating in the engine bed. The piston block lies slightly forward. The site is protected under the Historic Shipwrecks Act.

The iron screw-steamer *Georgette* (Official Number 10755) was built in 1872 at Dumbarton by McKellar and McMillan.[9] Built as collier, she had a capacity of 460 tons deadweight. Her two compound vertical double-acting steam engines produced 48 horsepower. The *Georgette* measured 46.2 metres long by 6.9 metres broad by 3.4 metres deep, and she carried two masts and schooner rig, a round stern, one deck with a raised quarter deck and a deck house. The vessel was given the classification of 90A1.

NOTES

1. *Inquirer*, 30 July 1873.
2. *Inquirer*, 27 December 1876.
3. 'Survivors Story', *West Australian*, 1 December 1935.
4. *Inquirer*, 31 January 1877.
5. Inquiry into stranding of the *Georgette*, Busselton, 21 December 1876, C.S.R. 824 (no folio numbers).
6. *Inquirer*, 10 January 1877.
7. *Inquirer*, 24 January 1877.
8. Report of inspection, Western Australian Museum file.
9. Lloyds Survey Register, National Maritime Museum, England.

Start

A letter to the *Inquirer* in March 1877 reported that a small cutter named the *Start*, belonging to a Mr McKay, had been totally wrecked on the Lacepede Islands prior to 27 January.[1] (See map on p. 277.)

A 6-ton cutter of that name (Official Number 61110) was built at Perth in 1873. That vessel had the dimensions 11 metres by 3.2 metres by 1 metre, and was built with one deck and a round stern. Her listed owners, however, were James and Ephraim Clarke, Thomas Hamilton and Henry Albert.[2]

NOTES

1. *Inquirer*, 21 March 1877.
2. Register of British Ships, Fremantle.

Aboyne, Isabellas, Helen Malcolm, Albert Victor, Three Lighters, One Ship's Boat, Cingalee and Mary Smith

During the 1850s and 60s, enterprising sea captains and shipowners began to exploit previously ignored phosphate deposits on islands in the South Seas. American vessels, in particular, searched for unoccupied guano islands and then set one or two men ashore, claiming that their law thereby entitled them to possession of the island's resources.[1] Understandably, this led to a painful new awareness within the governments of the Australian colonies of the vastness of the unoccupied sections of coastline and islands to be administered. British legal opinion was that the governments of the Australian colonies could not annex offshore rocks and islands merely by declaring an intent to annex them. It was necessary to occupy a place and preserve the evidence of that occupation.[2]

By the mid-1870s, the Western Australian Government had begun to take action. They had no alternative: an American named Mr S.P. Lord had planted his nation's flag on the guano-rich Lacepede Islands (several low, flat, rocky outcrops north of Broome), and claimed them for America.[3] And when Lieutenant Tooker of the joint Admiralty and Colonial Survey, with his party of surveyors, visited Ashmore Shoal, he found an American in possession.[4] Tooker hoisted the British flag and left two of his men in charge of Ashmore Shoal, while the Western Australian Government stationed an official representative on the Lacepedes, and required vessels to procure a licence before removing guano. This measure was designed to ensure that the Colony maintained control and received royalties for the guano. But of ten vessels anchored to collect guano at the Lacepedes in mid-February 1877, only three had the necessary licence.[5]

On the morning of the 16th, some of the vessels were loaded, and others partly so or in ballast. The dawn brought constant rain and a gale from east-north-east, producing huge breakers on the beach. The barques *Star of Jamaica* and *Amur*, and the brig *Emily*, slipped their cables and put out to sea, their masters fearful of the consequences of going aground on a lee shore with a tidal range of some 6 metres.

Three lighters, owned by the lessees of the guano deposits (Messrs Poole, Picken and Co., of Melbourne) and used for carrying guano out to the ships, were driven ashore. As the tide came in, the sea rose so much that the crests of the breakers were level with the height of the middle island, a terrifying spectacle for the twenty-seven men living ashore there, who had been employed by the lessees to collect and load the guano.[6] Soon after, the water flowed over the edge of the island, rolling forward planks and spars. The landing stages longing to the W.A. Guano Co. and to the barque *Amur* were washed away, together with bagged guano and a boat left behind by the *Amur*.

The next day, the barometer fell to 754 millimetres. At high tide, the waves flooded the camp and there appeared some danger that the whole of the low sandy island would be engulfed. At 1 p.m., when the tide was high, the master of the barque *Cingalee*, judging the time to be right, deliberately slipped his mooring and steered straight for the beach on the northern side of the western island, in an endeavour to save the lives of his crew. Being in ballast, the *Cingalee* struck lightly, and was lifted by the waves nearly on top of the island. All the crew scrambled ashore.

Closely following the *Cingalee* was the barque *Aboyne*, laden with 400 tonnes of guano. The *Aboyne*, with her deeper draft, struck further out to sea, about 0.4 kilometres eastward of the *Cingalee*. She fell, decks to leeward, and the sea quickly found an entrance to her hull. The guano from the hold was driven up in huge brown columns above her mast-heads.[7] The hull parted lengthwise, and in a few seconds, not a vestige was to be seen. Six lives were lost, including the captain's wife and two children.

The *Aboyne* had scarcely touched the beach when the schooner *Mary Smith* came ashore on a sandy point forming the brink of the narrow channel between two of the islands. The *Mary Smith* was left nearly dry by the receding wave, and her crew disembarked safely.

The brig *Isabellas* came ashore, followed by the barque *Helen Malcolm*, which struck close to where the *Aboyne* had disappeared. The *Helen Malcolm*'s masts quickly went over the side and the vessel suddenly collapsed.

The *Prince Arthur* and the *Albert Victor* were now the only vessels remaining afloat. The *Albert Victor*'s master signalled to the land party, 'Worst over and I think we will ride it out'.[8] Unfortunately, the winds increased again that

Wrecks litter the beach on the Lacepedes in 1877.

night. An enormous sea struck the barque *Albert Victor*, lifting her almost perpendicularly, and at that moment, her heaviest chain cable snapped. Captain McWilliam held on to his remaining anchors until high-water at 2 a.m., when he too ran for the beach. The *Albert Victor* went ashore close to the schooner *Mary Smith*.

The men living ashore on the middle island could observe, but not help, the men scrambling ashore on the western island as the vessels hit the beach. The Government Resident wrote:

> It was a painful sight to see the shore, as far as the eye could reach, strewed with the debris of the wrecks of the *Aboyne* and *Helen Malcolm*. Huge timbers broken in pieces, masts and other spars snapt across in short lengths of 6 feet and upwards. Planking literally reduced to matchwood. Copper bolts cut asunder, great iron bars twisted and turned like so much wire. Copper sheathing wrinkled like tin foil. Cabin fittings, boxes, clothing, cordage, sails, timber, iron tanks, all mingled together, broken, torn, smashed beyond all power of utilisation.[9]

The *Prince Arthur* rode out the storm, and later in the month, took sixty-four of the shipwrecked crewmen to Fremantle. The crew of the *Mary Smith* stayed on the island because the captain hoped to be able to refloat her. The vessel did float on the first spring tide, but drifted back onto rocks on the middle island while the crew were trying to get her ashore to be laid on her side for inspection of hull damage. So the crew took passage to Fremantle on another vessel early in March.[10] (See map on p. 277.)

The wrecks were auctioned in Fremantle. The *Cingalee*, the newest vessel, was knocked down for £170 to Messrs J. and W. Bateman. The remaining five vessels (and the schooner *Bessie* which had gone ashore at Beagle Bay during the same storm) were purchased by Captain Owston and partners for £340.

Soon after the sale, the news arrived in Fremantle that:

> The *Cingalee* which was worth at least £3,000 to a purchaser has been literally stripped of her gear (by pilferers) and could not possibly, even if successfully launched, be taken to any place of safety without a new outfit.[11]

Bateman and Owston sent the coaster *Azelia* to the Lacepedes with twenty workmen to break up the wrecked vessels, and others worked on salvaging the stranded ships. The *Cingalee* was the first vessel successfully refloated, and she arrived at Fremantle in September. The *Mary Smith* followed in October. These two vessels were repaired by the Fremantle shipwright Robert Howson and put back into useful service. The *Cingalee* was eventually lost when driven ashore at Bunbury in 1887, while the *Mary Smith* was beached and abandoned near Fremantle in 1890.

The remaining four vessels—*Aboyne*, *Isabellas*, *Helen Malcolm* and *Albert Victor*—were totally lost.

The 445-ton *Aboyne* (Official Number 48681) was built at Sunderland, in the United Kingdom, in 1864. She was copper fastened and was last yellow-metalled in 1875. The *Aboyne* was owned by W. Moodie and Co., and registered at London.[12] Her master was Captain Swan.

The 268-ton *Isabellas* (Official Number 28336) was built at South Shields, in the United Kingdom, in 1860. The *Isabellas* was fastened partly with iron bolts and was yellow-metalled in 1875.[13] She was owned by Thomas Brooks of Newcastle, New South Wales, and her master was Captain Friend.

The 311-ton *Helen Malcolm* (Official Number 54239) was built at Prince Edward Island, in Canada, in 1866. She too was partly fastened with iron bolts, and was yellow-metalled in 1869. The *Helen Malcolm* was owned by John Smith of Melbourne, and her master was Captain Kirby.

The 384-ton *Albert Victor* (Official Number 52835) was built at Schleswig, in Germany, in 1864.[14] She was owned by John Hughes of Williamstown, Victoria, and her master was Captain McWilliam.

Not one of the above wrecks has been found in modern times. The boat from the *Amur*, and the three lighters, described in the *Western Australian Times* as 'paltry old boats', would have disintegrated rapidly, while the *Aboyne* and *Helen Malcolm* broke up immediately. But what of the 268-ton *Isabellas* and the 384-ton *Albert Victor*, both of which went ashore at high tide? It would appear that the work begun by pilferers living on the island, and continued by Bateman and Owston's workmen, was efficiently completed by the heavy tides.

NOTES

1. Messrs Comlees and Daldy to Commodore Seymour, Auckland, 1 November 1861, Foreign Office 97/39, Public Record Office, London.
2. Lord Cairns to Lord Carnarvon, 3 May 1877, Foreign Office Correspondence, Public Record Office, London.
3. *Inquirer*, 8 August 1877.
4. *Western Australian Times*, 20 December 1878.
5. Richard Wynne (Officer in Charge, Lacepedes) to Col. Sec., 13 February 1877, C.S.R. 863, fol. 176.
6. Wynne to Col. Sec., 23 February 1877, C.S.R. 864, fol. 26–33.
7. *Inquirer*, 28 March 1877.
8. *Western Australian Times*, 13 April 1877.
9. Wynne to Col. Sec., 23 February 1877, C.S.R. 864, fol. 26–33.
10. Wynne to Col. Sec., 3 March 1877, C.S.R. 864, fol. 42.
11. Wynne to Col. Sec., 5 April 1877, C.S.R. 864, fol. 65.
12. *Lloyds Shipping Register*, 1877–78.
13. *Ibid*.
14. Board of Trade Wreck Register, 1877.

Bessie

The 228-ton schooner *Bessie* was stranded at Beagle Bay on 18 February 1877 during the same storm that claimed the *Aboyne* and other vessels on the Lacepedes.

The *Prince Arthur*, the only vessel to ride out the storm on the Lacepedes, cleared off for Beagle Bay on 25 February to procure water. On the way across to the mainland, the *Prince Arthur* fell in with a boat containing Captain Harrison and eight men from the *Bessie*. Captain Harrison explained that a cyclone from the north-north-east on the 17th had dragged the *Bessie* across a reef, starting a leak and resulting in the loss of her false keel and rudder.

With constant pumping, the *Bessie* was kept afloat until the next day, but she was striking heavily at low-water, so when the tide made sufficiently Captain Harrison slipped his cables and ran for the beach.[1]

The *Prince Arthur* stood into the Bay, but was forced by heavy seas to lie 10 kilometres off shore. The crew found the *Bessie* high and dry among the mangroves. They took water and provisions from the hulk, and returned to the Lacepedes. Lloyds listed the *Bessie* as 'abandoned'.[2]

The *Bessie* was later salvaged, along with the *Cingalee* and *Mary Smith*, and arrived at Fremantle, under the command of Robert Owen, in September 1877. The vessel was eventually wrecked and burned at Surabaya, in Java, in 1890.[3]

NOTES

1. Richard Wynne (Officer in Charge, Lacepede Islands) to Col. Sec., 28 February 1877, C.S.R. 864, fol. 40.
2. *Lloyds Shipping Register*, 1877-78.
3. *Inquirer*, 16 April 1890.

Painting of the *Bessie* in Asian waters.

Harrison

At the beginning of 1877, the firm of Connor and McKay, which had earlier brought the ill-fated steamer *Georgette* to Western Australia, was engaged in the construction of a jetty at Port Lyttelton, in New Zealand. The firm chartered, from Captain Harrison Godfrey in Adelaide, the three-masted schooner *Harrison* to take a cargo of jarrah from Bunbury to Port Lyttelton. The ships used for transporting timber in the latter years of the nineteenth century were frequently notoriously unseaworthy, and the *Harrison* proved to fit this description.

The *Harrison* arrived at Bunbury in March, under the command of Captain Godfrey. It was loaded with about £1,000 worth of jarrah, which Connor estimated would realize £2,387 in Lyttelton. At Bunbury, Connor was disturbed to hear Captain Godfrey say that the *Harrison* was 'a worthless old tub, which in all probability would never reach her destination'.[1] The vessel was, Godfrey explained, 40 or 50 years old, had previously been dismantled and condemned, and had recently been bought at San Francisco for a mere £300 before being laden with lumber and sent to Adelaide for sale. Notwithstanding this information, Connor allowed his timber to be shipped in her because she was insured!

So it was that the *Harrison* set sail from Bunbury on the morning of 23 March, bound for Lyttelton, with a light breeze and fine weather. Soon after leaving port, the vessel began leaking. The leak increased to such a rate as to compel the vessel to put back before she had rounded Cape Leeuwin. By then, she was making between 15 and 25 centimetres an hour, so Captain Godfrey decided to make for Fremantle. On his way, he ran the vessel, in broad daylight and fair weather, onto the Murray Reef. The keel was damaged, but she floated off after 3 or 4 hours and continued on to Fremantle.

On examination at Careening Bay, it was found that the fore keel was off, there were seven planks decayed on the starboard side, the caulking was out on the bow port side, and there was other damage. It was estimated that the repairs would cost about £1,100. Captain Godfrey decided that it would be cheaper to condemn and sell the vessel than to incur the expense of having her repaired.

The *Harrison* was condemned, dismantled and sold, and Connor claimed damages for the non-delivery of his goods. The court held that the vessel was unseaworthy when she left Bunbury, and granted Connor £700 damages.[2] Captain Godfrey apparently fell upon hard times and was reported to be in prison in Adelaide in 1878.[3]

Little is known of the subsequent fate of the *Harrison* hulk. In August 1881, James Lilly and Co. (agents for Anderson and Marshall, who ran the Government-subsidized steamship service) called for tenders to pump out, float and securely anchor the hull, then lying stranded at Keys Point, near the Rockingham Landing.[4] Then in April 1882, the *Herald* reported that 'the coaling hulk *Harrison* has been towed over to Careening Bay, Garden Island, in order to take in coals from the *Italy*'.[5] It seems likely, given the background of the vessel, that it would soon have proved unserviceable even as a coal hulk. It would appear that the hulk was bought, along with others, by the Adelaide Steamship Company in 1883.[6] A coal hulk uncovered on the seabed in the north end of Careening Bay during dredging work in 1973 yielded samples of the conifer Pinus ponderosa and the hardwood black locust, both American species.[7] However, it is known that a number of the coal hulks used in the Bay were American-built vessels. (See map on p. 242.)

The 384-ton three-masted schooner (or barquentine) was American registered. She carried eleven crew, and, on arrival at Fremantle, had a draft of 4.9 metres.

NOTES

1. *Western Australian Times*, 7 December 1877.
2. *Western Australian Times*, 21 December 1877.
3. *Inquirer*, 3 April 1878.
4. *Inquirer*, 17 August 1881.
5. *Herald*, 14 April 1882, p. 2.
6. *The Log*, 31 July 1968, p. 87.
7. Personal communication from D. E. Weiss (CSIRO Forests Products Laboratory, South Melbourne) to G. Henderson, 2 April 1973.

Hope

The 22-ton fore-and-aft schooner *Hope* (Official Number 72474) left Albany for Fremantle in April 1877 and was not seen again.[1] The Register entry was closed in 1878. The vessel had been built in Fremantle in 1875, with one deck, two masts, an oval stern and the dimensions 16.2 metres by 4.1 metres by 1.7 metres. Her original owners were William Dalgety Moore and John Finnerty. (See map on p. 164.)

NOTE

1. Register of British Ships, Fremantle.

Hadda

The barque *Hadda* arrived at the Lacepedes in April 1877 from Melbourne. Like the ill-fated *Aboyne* and other vessels, the *Hadda* had been chartered by Poole, Picken and Co., the lessees of the islands. But Captain Parker was told that he would have to procure the necessary licence from the Commissioner of Crown Lands in Perth before he would be allowed to load his vessel. So the *Hadda* sailed in ballast for Fremantle.

The course steered was calculated to miss the Abrolhos altogether, but the *Hadda* was steering very badly, the helm being constantly adjusted. The vessel was heading south-east by south when she struck on the Abrolhos at 10 p.m. on 30 April. She struck forward and ran right up on the reef.[1] Immediately, the water began to rise in the hold, so the sails were furled and the boats taken out with anchors, to attempt to warp the vessel off the reef. Such efforts were to no avail. (See map on p. 44.)

In the morning, the eleven crew would have seen Beacon Island close by, and it seems probable that they would have explored the island. They remained on board for their accommodation until 7 May, by which time the water was up to the *Hadda*'s lower deck beams. Then they set out in two boats for Geraldton, arriving the next day.

Museum maritime archaeologist Graeme Henderson supervises a student from the Diploma Course in Maritime Archaeology who is examining the *Hadda* remains.

A court of inquiry at Geraldton found that the vessel was wrecked due to a strong current pulling it off course. The wreck was auctioned in Geraldton, and Captain Parker bought the hull for £150.[2]

In recent times, Beacon Island was the base camp for the Western Australian Museum's excavation work on the *Batavia* wreck. The wreck of the *Hadda* lies some 700 metres east-north-east of Beacon Island in latitude 28° 25.4′ south, longitude 113° 47.5′ east. The site is protected under the Historic Shipwrecks Act. In 1980, students doing fieldwork for a post-graduate course in maritime archaeology did a detailed survey and limited excavation of the wreck site under the supervision of Museum archaeologists, showing that some of the wooden hull had survived in the coral waters.[3] A small rectangular stone structure on the south-east point of Beacon Island has been found to contain artefacts dated to the latter half of the nineteenth century, and it is thought that this may represent either a temporary campsite for the *Hadda*'s crew immediately after the vessel was wrecked, or a camp for some later salvors.[4]

The 334-ton *Hadda* (Official Number 61014) was built in Germany in 1860.[5] She was owned by William Johnston, of Newcastle, New South Wales.

NOTES

1. Evidence of Robert Crickmore, First Officer, 11 May 1877, Police Records, Acc. No. 129, 24/714, Battye Library.
2. *Inquirer*, 16 May 1877, p. 2.
3. L. Vickery and I. Spooner, A Report on the 1980 Excavation of the *Hadda* Site in Houtman Abrolhos (Western Australian Museum Library).
4. B. Bevaqua, Archaeological Survey of Sites Relating to the *Batavia* Shipwreck, May 1974 (Western Australian Museum Library).
5. *Mercantile Navy List*, 1877.

Hampton

The 19-ton cutter *Hampton* was abandoned and then stranded at Port Grey, just south of Geraldton, on 11 May 1877. The vessel had left Fremantle on the 4th, bound for Shark Bay. The following evening, a gale tore away the lower and middle gudgeons of the rudder. A jury rudder system was tried, and Captain Outred made for Champion Bay. But on the evening of the 11th, the *Hampton* drifted among breakers and started bumping. The four crew got into the dinghy and soon lost sight of the cutter in a squall. They made the land, about 16 kilometres north of Geraldton, the following day. Meanwhile, the *Hampton* drifted ashore, where it lay undamaged, apart from the missing rudder.[1]

The vessel was refloated two weeks later, and all of the cargo was saved.[2] Captain Outred then continued on to Shark Bay.

NOTES

1. *Inquirer*, 13 June 1877.
2. *Inquirer*, 23 May 1877.

Eulie

The 335-ton barque *Eulie* was stranded on the outlying rocks known as the Horseshoe Reef, at the West End of Rottnest Island, on the evening of 16 May 1877.

Rhodes, in his *Pageant of the Pacific*, states that the *Eulie* was wrecked near Rottnest, but in fact it was a relatively minor casualty.[1] The *Eulie* bumped heavily three times, and on arrival at Owens Anchorage, was making 10 to 13 centimetres of water an hour.[2] The *Eulie* was repaired at Careening Bay and later departed for Singapore with 200 tons of sandalwood.

NOTES

1. F.C. Rhodes, *Pageant of the Pacific: Being the Maritime History of Australia* (Sydney, 1937).
2. Minutes of Inquiry into the Stranding of the *Eulie*, C.S.R. 865, fol. 236–247.

Twilight and Bunyip

On the night of 24 May 1877, the cutters *Twilight* and *Bunyip* were discharging stores from Albany intended for the parties engaged in the construction of the Eucla telegraph line, when they were driven on shore in a gale.[1] No lives were lost. Both vessels were abandoned as wrecks at the eastern end of Culver Cliffs, at a place subsequntly named Twilight Cove. One source has it that the *Bunyip* went on shore about 180 metres east of the cliffs at low tide, while the *Twilight* went ashore another 180 metres east at high tide.[2]

The crews of the two vessels scrambled ashore safely and walked overland some 275 kilometres to Israelite Bay to get help.[3] In 1900, a camper at Twilight Cove reported one wreck to be lying 0.4 kilometres inland. He and his mates used wood from the wreck to build a well.

The *Twilight* was the same vessel that had been stranded at Bunbury with the *Midas* in 1872. The wreck now lies partly buried in sand some distance inland from the shoreline at Twilight Cove. The 24-ton *Twilight* (Official Number 61091) was built at Fremantle in 1869, with one deck, one mast and a

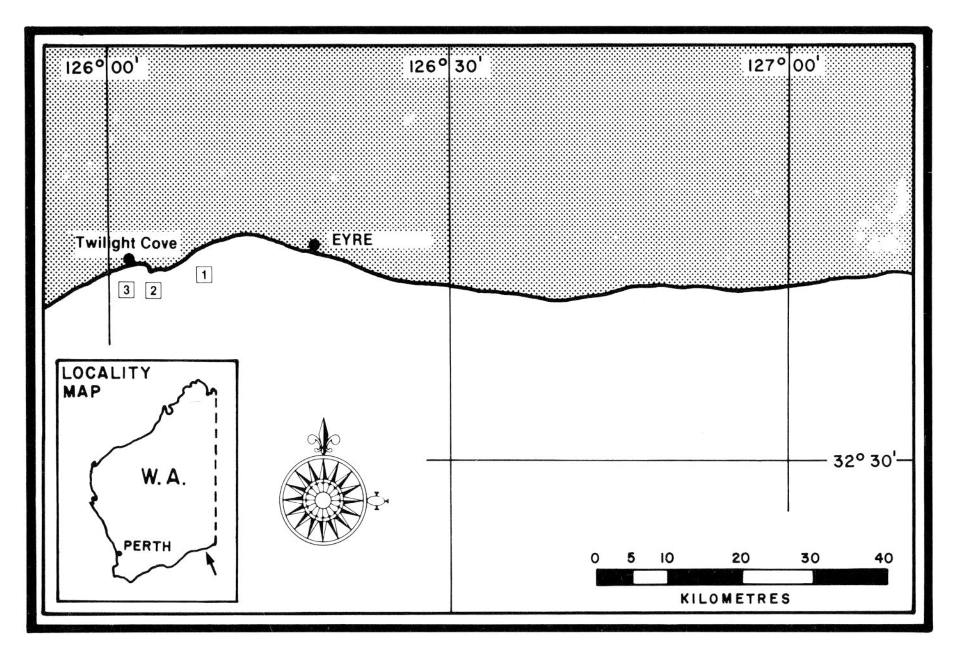

1 UNIDENTIFIED WRECK PRE-1841
2 *TWILIGHT*
3 *Bunyip*

square stern.[4] Her dimensions were 13.1 metres by 4.3 metres by 2 metres. She was owned by George Thompson, of Fremantle.

One account states that the *Bunyip* was built at King River, Albany, and named after the hill beside which she was launched. A vessel of the same name was registered at Sydney as number 69 of 1873. The 57-ton Sydney vessel (Official Number 64443), built of wood with a steam engine, was still afloat in 1887.[5]

NOTES

1. L.N. Clifton to Col. Sec., 25 March 1879, C.S.R. 956, fol. 79.
2. W.P. Prideaux, King River, 15 February 1922, to anonymous, Western Australian Museum file.
3. G.P. Stevens, 'The East-West Telegraph, 1875–77', *R.W.A.H.S.J.*, XIII, p. 3
4. Register of British Ships, Fremantle.
5. Register of British Ships, Sydney.

Conservator Neil North measures a part of the *Twilight* wreck from the dunes at Twilight Cove.

Belle

The brigantine *Belle*, laden with jarrah timber for Port Adelaide, was driven ashore at Lockeville in Geographe Bay during a gale on 20 August 1877.[1] However, the vessel was soon refloated, found not to have sustained serious damage, and reloaded.

NOTE

1. Inquiry into stranding of brigantine *Belle*, 21 August 1877, C.S.R. 904, fol. 91.

Gertrude

The Government's revenue cutter *Gertrude* suffered two accidents on a visit to the Lacepedes from Cossack in January 1878. *En route*, the master, Pemberton Walcott, called in at a previously unknown anchorage south of Depuch Island—later named Balla Balla Harbour. While there, however, he lost his best bower anchor with 27 metres of chain, when it became fouled on a rock.[1]

When the *Gertrude* arrived at the Lacepedes and anchored, the weather deteriorated. On 19 January, it reached cyclone force from north-north-east, and the *Gertrude* dragged onto a sandy beach. Two days later, the captain of the *Prince Arthur* made a heavy anchor available so the *Gertrude* could be heaved off the beach with little damage.

NOTE

1. Pemberton Walcott to Col. Sec., Cossack, 11 February 1878, C.S.R. 885, fol. 2.

Marten

The 27-ton schooner *Marten* was built on the bank of the Yarra, in Victoria, in 1871. Her original owner, shipbuilder Robert Kennedy, sold her in 1874 to Fremantle master mariner George Robinson.[1] While in Western Australia, she was engaged in the pearling industry at Cossack. In May 1876, she made an exceptionally fast 2-week voyage under canvas from Timor to Fremantle.[2] In the 1870s, the accepted sailing average from Fremantle to the Malay Archipelago was 25 days.

In January 1878, the *Marten* was bought by the Browse Island Guano Company, of Adelaide, who intended to use the tiny craft as a dispatch boat to keep up communication between Browse Island and Darwin. She called through Fremantle at the beginning of March 1878, on her way from Adelaide to Browse Island, carrying ballast and stores, and a medical man for the workers on the island.

After leaving Fremantle, Captain Alexander Dewar steered north-north-west before strong south-east winds, and expected to pass 40 kilometres westward of the Abrolhos. But just before 11 p.m. on 4 March, breakers were seen a kilometre or so ahead. Captain Dewar later explained:

> We set the foresail, trimmed the rest of her sails and hauled the ship by the wind. She missed stays when I attempted to head her round and finally got into the breakers. She was a vessel that would not steer well at all and failed several times before . . . The vessel was running dead before the wind and I brought her to the wind on the port tack, the boom being on the starboard side. Next morning I found the vessel was ashore right on the middle of the South Island, Pelsart Group on S.W. side. The schooner went broadside onto the breakers, when she struck she remained on the rock for about 10 or 15 minutes. The false and main keel, rudder and starboard bilge of the vessel are gone.[3]

The *Marten* was holed, and soon filled with water. Captain Dewar and his five crew got ashore on Pelsart Island and rescued some of the stores. However, it was clear that the *Marten* could not be refloated without repairs, and that even if she was repaired, it would be very difficult to get her back to sea through the breakers (the island is bounded by a shallow reef platform inside the breakers).

Captain Dewar left three of the crew in charge of the wreck, and set out for the mainland, with two of his crew, in an open boat. They arrived at Geraldton on the 13th.[4] Dewar hoped to be able to save the *Marten*, but there is no record of the vessel being refloated. (See map on p. 44.)

The *Marten* (Official Number 64770) had the dimensions 15.4 metres by 4.7 metres by 1.7 metres. She was built with one deck, two masts, a billet head and a round stern.[5]

The site of the *Marten*, at latitude 28° 55.5′ south, longitude 113° 59.6′ east, is protected under the Historic Shipwrecks Act. Today, nothing can be seen of the hull, but several objects found by the Western Australian Museum on Pelsart Island may be associated with the wreck.

NOTES

1. Register of British Ships, Melbourne.
2. *Inquirer*, 31 May 1876.
3. Alexander Dewar, evidence at inquiry into the loss of the *Marten*, 13 March 1878, C.S.R. 885, fol. 84.
4. *Inquirer*, 13 March 1876.
5. Register of British Ships, Port Adelaide.

Young Victorian, Chip, Ethel, Venus, Mystery and Emma

In March 1878, a destructive cyclone visited the North West, resulting in loss of life and damage to property. On the morning of 4 March, the pearling vessels *Venus*, *Emma*, *Ethel*, *Chip* and *Young Victorian* were at anchor at Forestier Islands, to the east of Cossack. (See map on p. 32.)

The *Young Victorian*, a small craft of less than 5 tons that belonged to David Stewart, was blown out to sea, and it was supposed that she had struck a bank and capsized. She had on board two Europeans—Hugh McKellar and John Jones—and a Swan River Aborigine with his wife and child.[1] The Aboriginal woman was carried on the vessel in contravention of the labour laws.

The *Chip*, another small craft, owned by S. Isbester, was blown out of the water, capsized and dismasted. Isbester's body was found beneath the vessel.[2]

The 14-ton *Ethel* collided with the *Venus*, and sank immediately, her side stove in and her head-sails blown away. A boy named Blurton drowned in the cabin. The master, A. Rouse, could not swim, but was saved from drowning by his wife, who supported him for about a kilometre until they reached the shore.[3]

The *Venus*, a 14-ton vessel, lost her bulwarks and covering board, started all the planking on her bilge, and lost her jib and mainsail. The 17-ton *Mystery*, which was also in the area, lost her jib and mainsail. An Aboriginal boy named 'The Eye' jumped overboard to swim ashore and was drowned.[4] The 8-ton *Emma* had her masts cut away to prevent her capsizing.

Seven lives were lost in the disaster. The *Young Victorian* was found, bottom up, on Middle Forestier Island by Pemberton Walcott, searching in the revenue vessel *Gertrude*. The wreck lay in a dangerous exposed position and had broken up too much for any bodies to remain in the hull, so Walcott did not examine it closely.[5]

It seems likely that the other craft, with the possible exception of the *Chip*, were put back into service. The *Ethel* was again reported wrecked, this time to the west of the Fortescue River, in February 1881.[6]

No details of the *Young Victorian* have been found, and it is likely that she was not registered.

NOTES

1. Government Resident, Roebourne, to Col. Sec., 14 March 1878, C.S.R. 897, fol. 124.
2. *Inquirer*, 17 April 1878.
3. Government Resident, Roebourne, to Col. Sec., 14 March 1878, C.S.R. 897.
4. Government Resident, Roebourne, to Col. Sec., 14 March 1878.
5. Pemberton Walcott to Col. Sec., 20 April 1878, C.S.R. 885, fol. 15.
6. *West Australian*, 8 February 1881.

Carleton and Matterhorn

Another cyclone struck the North West just a week after the March 1878 disaster at Forestier Islands. Messrs De Beer and Co.'s Melbourne monthly shipping report related:

> We regret having to report the loss of two vessels at Browse Island in a hurricane which passed over on 11 March. The German ship *Matterhorn* 1,306 [tons], with loss of captain and 16 hands, together with 4 of the crew of the German barque *Flora*, who went to their assistance; and the British barque *Carleton*, the crew of the latter being saved. The information comes forward by wire and we fear further losses will be reported thence, as one or two vessels are overdue.[1]

The disaster was very substantial—twenty-one lives lost, together with over 2,000 tons of shipping—yet it was hardly noticed by the Perth press. A confusing situation arose in March 1879, when one Perth newspaper announced that a cyclone during the latter part of December 1878 had struck Browse Island and wrecked 'the whaler *Runnymede*'. Another Perth newspaper announced that the same cyclone wrecked several vessels at Browse Island and Ashmore Shoal, including 'the *Matterhorn* (American); the *Runnymede* and the *Selina* (both belonging to Liverpool); and the *Carleton* (British)—all large vessels of 1,000 tons and upwards—were lost at Browse Island'.[2] (See map on p. 152.)

It seems probable that both the *Carleton* and the *Matterhorn* were indeed wrecked in March 1878.

The Board of Trade Wreck Register gives the date of 10 March 1878 for the total loss of the *Carleton* (Official Number 61797).[3] That vessel was built in 1870 at Beaver River, New England, in the United States by G. Jenkin.[4] It was a 742-ton barque with the dimensions 49.1 metres by 10.8 metres by 6.1 metres, owned by G. Doane and Co., and registered at Yarmouth. The *Carleton* was built of spruce, pine, birch, hackmatack and oak.

The 1,306-ton three-masted *Matterhorn* was built at Bath, Maine, by E. and A. Sewall in 1866. The vessel measured 57.3 metres by 11.6 metres by 7.3 metres, was owned by A. H. W. Waffaus and was registered at Hamburg.[5] She was built of oak.

NOTES

1. De Beers Monthly Shipping Report, quoted in *Inquirer*, 1 May 1878.
2. *Inquirer*, 19 March 1879; and *Western Australian Times*, 25 March 1879. See also *Herald*, 22 March 1879
3. Board of Trade Wreck Register, National Maritime Museum, England.
4. *Registre Veritas*, 1878.
5. *Ibid*.

Lady Elizabeth

During the 1870s, the barque *Lady Elizabeth* was regarded as one of the finest of the vessels employed in the trade between London and Fremantle. These vessels generally sailed direct from London to Fremantle, but on the homeward run, particularly if they missed the wool season, they frequently called at Chinese ports for more goods. The *Lady Elizabeth* left Fremantle at daylight on 25 June 1878, chartered by Messrs Shenton and Monger to carry a cargo of 611 tons of sandalwood to Shanghai. Captain Thomas Scott's daughter was the only passenger.

After sailing safely outside Rottnest Island, the *Lady Elizabeth* was driven southward by heavy weather, and continuing gales made it impossible for Captain Scott to take navigational observations for a number of days. On the morning of 30 June, Captain Scott decided to turn back to Fremantle. Rottnest Island was at that time bearing south-south-east by south about 55 kilometres away.[1] At 3.30 p.m., a man was lost overboard. Two lifebuoys were immediately cut loose and thrown overboard, but it was a hopeless gesture. The seas were too heavy to lower a boat to provide real help.

At about 4 p.m., Captain Scott sighted Rottnest to the north, and 4 hours later, he estimated the distance to be down to 8 or 10 kilometres away. At 8.30 p.m., the lookout saw land and breakers on the lee beam, so the helm was put up immediately on to a course of east-south-east. A few moments later, the *Lady Elizabeth* grazed the bottom, but passed on.

The barque *Lady Elizabeth*. This may be the replacement vessel, built in 1879.

Captain Scott saw what he judged to be Parker Point on the port bow and changed his course to north-east, intending to sail into Fremantle via the channel south of Rottnest Island. But after a short time on the new course, the barque suddenly struck heavily twice on the reef in Bickley Bay, knocking the man from the wheel and sending the ship's head round to the southward. The *Lady Elizabeth* was then unmanageable, so Captain Scott ordered the crew to let go the port anchor. (See map on p. 242.)

A short time later, the carpenter reported that there was 1.5 metres of water in the well, and by 10.30 p.m., the water was over the upper deck. The ship was gradually falling over to starboard. Captain Scott described the scene:

> Towards midnight terrific squalls of wind and rain with heavy sea, burst in, the front of the poop gutted and running half way along the poop on the after port corner of which were huddled together my daughter, officers and crew and self where I ordered them to remain till daylight.[2]

Captain Nash, the Pilot at Rottnest, saw the blue distress rockets early in the night and put to sea in his lugger, but could not approach the wreck until daylight, when he took the crew ashore. Two Fremantle boats approached the wreck during the day and found that sandalwood was already breaking free of the hold, but Captain Scott rowed himself alongside and threatened to prosecute anyone who touched the ship or the cargo. Nevertheless, the salvors and beachcombers had a very profitable time of it during the following days, as cargo was washed ashore from Rottnest Island to Bunbury.[3]

The hull, together with the lead ore cargo and the sandalwood, was sold at auction for a total of £1,039. [4] The sandalwood cargo, previously valued at £5,000, was insured.

A court of inquiry was held. Captain Scott, during the course of proceedings, 'made use of expressions . . . which were both unbecoming and amounted to gross contempt.'[5] Nevertheless, the court did not press specific charges of incompetence or misconduct.

The composite-built 658-ton barque *Lady Elizabeth* (Official Number 60966) was built in 1869 at Sunderland by Robert Thompson.[6] The keel consisted of American rock elm, with English elm at the fore end. The vessel's stem was of teak and English oak, her stern-post of teak, and her apron and floors of iron, while the planking outside was of American rock elm. She measured 48.7 metres by 9.3 metres by 5.5 metres, and she carried one deck. The *Lady Elizabeth* belonged to the well-known firm of Messrs Wilson and Oliver.

A year after the *Lady Elizabeth* was lost, Robert Thompson had built another vessel of the same name for John Wilson.

Today, the wreck lies in some 7 metres of water on the shoreward side of Dyers Island in latitude 32° 01.1′ south, longitude 115° 32.8′ east. Because of its sheltered position and relative structural integrity, the site has been used by members of the Maritime Archaeology Association to develop their survey skills underwater.

Recreational divers examine the information plaque on the *Lady Elizabeth* wreck.

NOTES

1. Captain Scott, evidence at inquiry, Fremantle, 17 July 1878, C.S.R. 885, fol. 152, Battye Library.
2. Captain Scott, evidence.
3. *Inquirer*, 31 July 1878.
4. Auction of 12 July 1878, Auction Book No. 312, Samson Papers, 1120A/20, Battye Library.
5. *Western Australian Times*, 16 August 1878.
6. Lloyds Survey Register No. 9627, National Maritime Museum, England.

Brothers

In January 1876, the *West Australian* reported on an addition to the local coasting fleet:

> On the 20th inst. Mr James Storey made another successful launch of a pretty model schooner of 50 tons burden. She is built of jarrah, the most durable wood known for ship building purposes, and effectively resists all attacks of marine insects. Her owners are Messrs H. Higham and Son who have had her built expressly for the pearl shell fishery trade and christened her the *Brothers*. It is to be hoped she meets with a better fate than the pioneer vessel of the same name fitted out by Messrs Francisco Brothers some nine years ago, the wreck of which is supposed to have been recently discovered by Mr. J. Tuckey on the North West Coast.[1]

Unfortunately, the vessel met the same fate as her namesake a little over 2 years after completion. Perhaps she would have been safer in the pearling industry than on the south coast:

> The *Brothers* was chartered for the purpose of taking Andrew (Dempster) and his family back to Esperance. But the little schooner passed into a storm and encountered mountainous seas. The captain sought solace in drink and the navigation was left to Andrew Dempster. The schooner leaked so badly that all hands were ordered to the pumps. Although Andrew's family are now men and women with grown up children of their own, none who were old enough to remember have forgotten that terrible voyage. They still recount with evident feeling their sensations in that wind hounded wave buffeted boat. The picture of a little schooner braving a big sea, with Andrew Dempster at the wheel, with five children and a wee baby aboard, with all hands at the pumps, is not one lightly forgotten. At last the barque rolled into Esperance. After the Dempsters disembarked, the *Brothers* put out to sea, but nothing has been heard of her from that time to this.[2]

The *Brothers* left Esperance for Fremantle on 12 July 1878. At the end of September, a schooner was seen anchored at Hamelin Bay, and in mid-October, a vessel was seen at Cape Naturaliste.[3] Neither proved to be the *Brothers*, which was never seen again. (See map on p. 164.)

The 48-ton topsail schooner *Brothers* (Official Number 72481) was built at Fremantle, with one deck, two masts, and an oval counter. Her dimensions were 19.7 metres by 5.2 metres by 2.2 metres.[4] She was owned by Mary Higham and Edward Higham, both of Fremantle.

NOTES

1. *Western Australian Times*, 25 January 1876.
2. 'The Brothers', *Muresk College Magazine*, September 1928.
3. *Inquirer*, 2 October 1878 and 16 October 1878.
4. Register of British Ships, Fremantle.

Diana

The three-masted schooner *Diana* arrived off Fremantle on 4 July 1878, carrying ballast from Port Natal. Captain Humphrey Humphreys sailed into Gage Roads without waiting for the services of a pilot, and suffered the consequences when he stranded the vessel on the Parmelia Bank.[1] With the assistance of the Harbour Master, the *Diana* was refloated without damage and anchored at Owens Anchorage.

On 15 July, a heavy gale struck the port. Captain Humphreys had the *Diana*'s royal yards taken down. The vessel was moored with 177 metres of chain on the starboard anchor and 69 metres on the port anchor. In a heavy squall at about 3 p.m. on 16 July, both cables broke and the *Diana* was swept onto the beach.[2] The coasters *Clarence Packet* and *Argo*, and the lighters *Will Watch* and *Myth*, also went ashore.

The *Diana*, full of water and with her back broken, was condemned as a wreck and sold at auction by Messrs L. A. Manning. The hull was purchased by a Mr McCleery for £85. A subsequent court of inquiry found that no blame could be attached to the master or crew.[3]

The 224-ton *Diana* (Official Number 28766) was built in 1860 at Teignmouth, in Devon, with the dimensions 33.6 metres by 7.2 metres by 5.5 metres. Her wooden hull was partly fastened with iron bolts and sheathed in yellow-metal, last replaced in 1876. She was registered at Aberystwyth, in Wales, and listed as owned by Mrs E. Edwards.[4]

An indication of the appearance of the wreck before the turn of the century is given in a letter, written in 1973, by Lucius Manning, then aged 93:

> When I was a kid there were two old ships in Owens Anchorage: the *Juno* [presumably the *James*, wrecked in 1830] and the *Diana*—I think they were whalers. The *Juno* was cannibalised. You could wade out to her but her timbers had been cut off down below water level.[5]

The implication would seem to be that the *Diana* had not been 'cannibalised', and that her remains protruded above the water. In 1975, the *Diana* site was relocated by skindivers George Green and Mike Pollard, of the Maritime Archaeology Association of Western Australia. The remains lie in latitude 32° 05.9′ south, longitude 115° 45.3′ east, about 75 metres south-east of the Fremantle Power Station's cooling water outlet, and are protected under the Commonwealth Historic Shipwrecks Act. (See map on p. 242.)

NOTES

1. *Inquirer*, 10 July 1878.
2. *Inquirer*, 31 July 1878.
3. Minutes of a Preliminary Court of Inquiry to inquire into the circumstances attending the stranding of the *Diana*, C.S.R. file 52/1878.
4. *Mercantile Navy List*, 1877.
5. Lucius Manning, notes left with the Western Australian Museum.

Ella Gladstone

The southern end of Geographe Bay was a dangerous place for sailing ships to anchor in winter. The *Governor Endicott* was blown ashore and wrecked in July 1840, the *Halcyon* in August 1844, and the *Geffrard* in June 1874. Several other vessels were stranded.

The brig *Ella Gladstone* was anchored at Quindalup taking on a cargo of timber for timber merchant Maurice Davies on 21 July 1878 when a north-westerly gale developed. The brig parted her cables during a violent squall, and despite the crew's efforts, the vessel was driven ashore. During the following week, she was condemned as a wreck by a board of survey.[1]

Auctioneers Lionel Samson and Co. advertised the wreck for auction, with the condition that Davies be given a reasonable time to take his timber out of

The brig *Ella Gladstone* with other craft at Port Adelaide around 1872.

the hull.[2] W. D. Moore purchased the hull for £52. An inquiry into the cause of stranding attached no blame to the captain and crew, nor to the ship's equipment.

In January 1879, the *Inquirer* carried the following note:

> The *Ella Gladstone* has been got off at Quindalup and is likely to turn out a good speculation for the purchaser.[3]

It is not known whether the vessel was repaired and re-registered, but it does not appear in *Lloyds Shipping Register* after 1878.[4] (See map on p. 204.)

The 225-ton brig *Ella Gladstone* (Official Number 28624) was built at Sunderland in 1860, with the dimensions 29.9 metres by 7.2 metres by 4.5 metres.[5] At the time of her wreck, she was owned by George Wilson, of Adelaide.

NOTES

1. *Inquirer*, 31 July 1878.
2. Lionel Samson and Co., Auction Books, 1878, Acc. 1120A, Battye Library.
3. *Inquirer*, 15 January 1879.
4. *Lloyds Shipping Register*, 1878, 1880.
5. Register of British Ships , Melbourne.

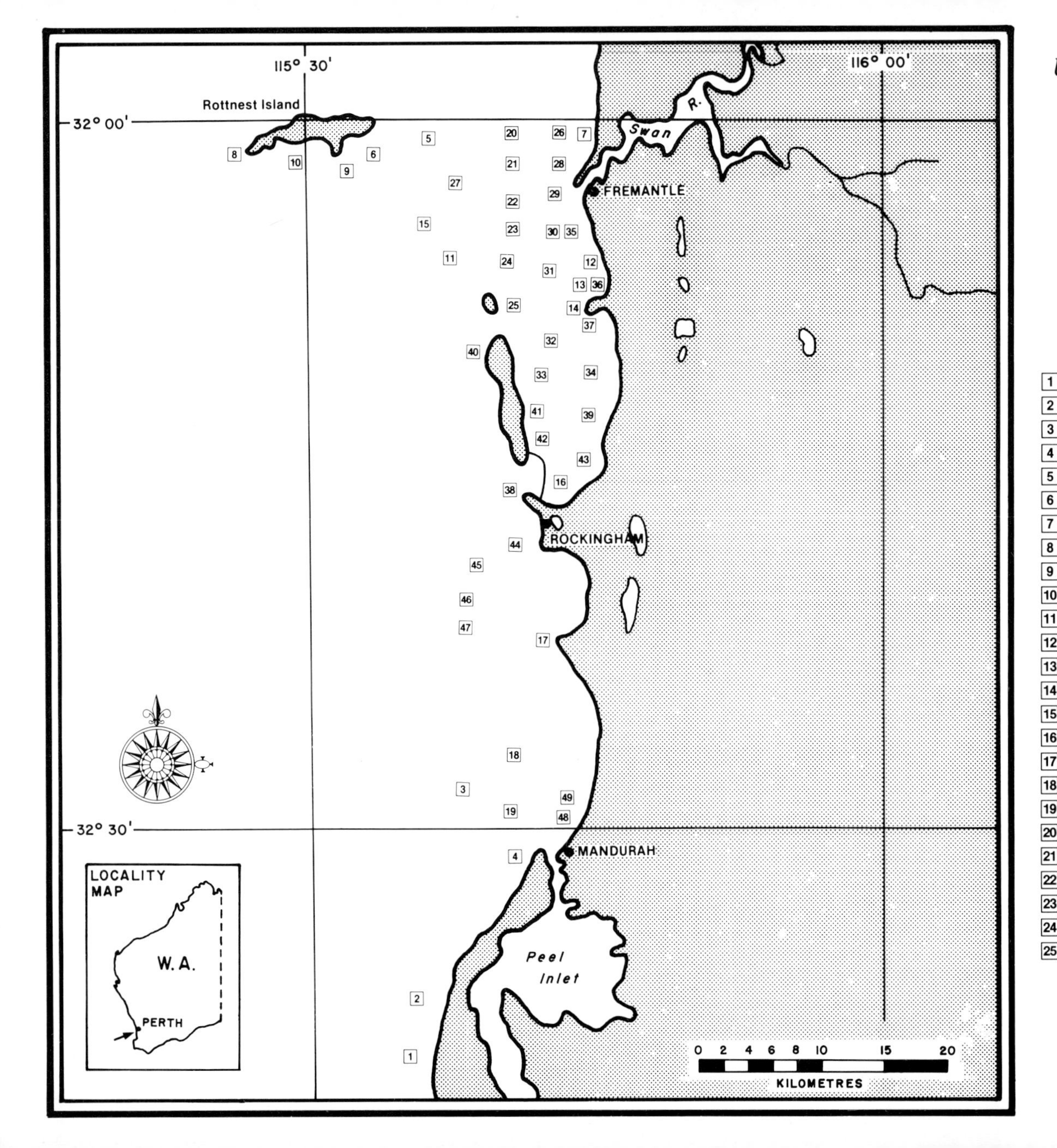

Unfinished Voyages 1851–1880

1 *Lass of Geraldton*
2 *Bee*
3 *Young Shepherd*
4 *Alert*
5 Pilot Boat 1851
6 *TRANSIT*
7 PS *Endeavour*
8 Ship's Boat 1875
9 *GEM*
10 *LADY ELIZABETH*
11 *LANCIER*
12 *JAMES*
13 *DIANA*
14 *JAMES MATTHEWS*
15 *ZEDORA*
16 *CONTEST*
17 *HERO OF THE NILE*
18 *CHALMERS*
19 *JAMES SERVICE*
20 *Wave*
21 *Waterwitch*
22 *Antelope*
23 *Inbat*
24 *Gazelle*
25 *Faith*
26 *Canning*
27 *Amelia*
28 *Emily Taylor*
29 *Thames*
30 *Marquis of Anglesea*
31 *Sea Nymph*
32 *Avon*
33 *Lord Stanley*
34 *Fanny*
35 *Fitzgerald*
36 *Vixen*
37 *Alma*
38 *Devonshire*
39 *Lady Stirling*
40 *Twinkling Star*
41 *Albion*
42 *Rockingham*
43 *Harrison*
44 *Cumberland*
45 *ROBERTINA*
46 *STAR*
47 *Bungaree*
48 *Black Swan*
49 *Preston*

James Service

The iron barque *James Service* sailed from Melbourne, under the command of Captain Young, late in 1877. The vessel called at Rockingham and took in jarrah sleepers for Calcutta. She left Calcutta on the return voyage to Melbourne on 23 April 1878, carrying in addition to her crew a theatrical party of ten and a cargo of 1,000 bales of sacks, 3,000 cases of castor oil, and 600 bales of jute.[1]

A short time after her departure, the vessel encountered problems. The *James Service* was becalmed for 16 days in the Bay of Bengal. In the intense heat, Captain Young suffered from severe sunstroke, which brought on fever, dysentery and delirium. The crew and passengers concluded that the Captain was insane, while he in turn became convinced that there was a conspiracy to depose him and put the mate in his place. The crew won the day, and Captain Young was put under restraint until the vessel reached Penang.[2] There, he was charged with being incompetent as a seaman, and with imperilling the vessel off the Nicobar Islands. The court found these charges not proved, but considered that his state of health made it inadvisable for him to resume command. The agents engaged Captain Sievwright to take the barque from Penang to Melbourne.

The *James Service* headed southward from Penang to round Cape Leeuwin, a course that should have taken her several hundred kilometres off the west coast of Australia. But she was wrecked on the coast on 22 July, probably during the early evening. The circumstances of the loss only gradually emerged. An Aborigine saw the masts above the surface of the water, west of the Murray River mouth, on the 23rd. Then a longboat bearing the words '*James Service*, Melbourne', was found on the beach. The following day, the shore was strewn with wreckage. One section of the beach was literally covered up to the high-water mark with thousands of cases and tins of castor oil. The first corpse was found soon afterwards. As more bodies were found, they were buried above the high-water mark.[3] A diary found washed ashore indicated that up to the 20th of July, the vessel had been encountering very boisterous weather for some time, and had been on her beam ends, the yards touching the water. Speculation as to why the vessel had been driven so far off course led to the theory that she had been storm damaged. There were no survivors. (See map on p. 242.)

Later in the year, a diver examined the hull and reported that the vessel had broken into two parts. The hull was sold at auction for a mere £20 in November 1878.

The wreck was rediscovered in latitude 32° 27.5′ south, longitude 115° 39.5′ east, lying in 6 metres of water on the Murray Reef, by members of the Underwater Explorers Club during a search in 1962 led by diver Harold Roberts. A 3-metre-long anchor was raised the following year, to be exhibited at Mandurah. Since that time, more skeletons have been found as winter gales shift the beach sands. The wreck itself is now protected under the Historic Shipwrecks Act.

The 441-ton *James Service* (Official Number 55609) was built in 1869 at Govan, County of Lanark, in Scotland, by Dobie and Co. Her dimensions were 46.8 metres by 8.6 metres by 4.7 metres and she carried three masts on two decks. The iron hull was clincher built, with a round stern and a demi--figurehead of a man.[4] At the time of her loss, she was owned by James Service, of Melbourne, and others.

Remains of one of the crew from the *James Service*. Skeletons are occasionally found in the sand hills adjacent to the wreck.

NOTES

1. *Herald*, 27 July 1878.
2. *Western Australian Times*, 20 August 1878.
3. *Inquirer*, 21 August 1878.
4. Lloyds Survey Register, No. 7518, National Maritime Museum, England. See also Register of British Ships, Melbourne.

Star Queen

On the 2nd August 1878, the *Western Australian Times* reported 'the total loss of the *Star Queen* from Melbourne . . .'.[1] But the newspaper was wrong. The barque had stranded on the Sisters Reef on the evening of 29 July, because her master, Captain Henry Shelton, had neglected to keep a check on the vessel's distance from shore.[2] The *Star Queen* lost her rudder on the reef, but after her cargo of wheat and flour was discharged, she drifted inside the reef towards Long Point, where she was anchored.

The Fremantle Harbour Master, George Forsyth, took the vessel to Careening Bay for examination. She was put up for auction in Careening Bay in February 1879, and bought by T. & H. Carter, of Fremantle. After her repairs were completed, the *Star Queen* did some trading out of Fremantle, including a trip to Singapore in 1881, and was sold at Port Louis in 1889.[3]

The 264-ton *Star Queen* (Official Number 51306) was built at Kingsbridge, Devon, in 1866, and was owned by Grant and Co., of Salcombe. The vessel's dimensions were 35.8 metres by 7.7 metres by 4.4 metres.[4]

NOTES

1. *Western Australian Times*, 2 August 1878.
2. *Western Australian Times*, 25 October 1878.
3. Register of British Ships, Fremantle.
4. *Lloyds Shipping Register*, 1878–79.

Sharperton

Frank Goldsmith states that a vessel named *Sharperton* was stranded off the eastern Lacepedes on 10 August 1878, and that the master, Captain Atkinson, was found, in a subsequent inquiry, to have lacked common caution in approaching a strange coast, to have failed to keep a lookout, and to have grounded his ship within limits shown on Admiralty Charts.[1]

Lloyds Shipping Register lists a vessel of that name (Official Number 60518), commanded by T. Atkinson.[2] This was a 363-ton barque, yellow-metalled in 1879, perhaps because of the 1878 stranding.

NOTES

1. Frank Goldsmith, *Treasure Lies Buried Here* (Perth, 1946), p. 219.
2. *Lloyds Shipping Register*, 1879.

Salve

The brigantine *Salve* was at Lockeville, loading jarrah, on 7 September 1878 when a south-westerly gale drove her ashore. Captain John Campbell later told an inquiry that the vessel started bumping on the seabed during the first part of the night, and after one chain cable broke, he decided to pay out the remaining cable and back her onto the beach.[1] He described the vessel as being very strongly built. The court found that Captain Campbell had committed an error of judgement in anchoring inshore on bad holding ground.

The jarrah cargo was discharged, and marine surveyors reported the hull to be in good order, so the *Salve* was refloated and reloaded, and she left for Wallaroo in South Australia on 24 September. The following day, the vessel was caught in a gale off Cape Leeuwin. The yard of the fore topsail was blown away and the vessel sprang a leak. In order to save the lives of his crew, Captain Campbell again deemed it prudent to run the vessel ashore. She was accordingly beached at Port Augusta, in Flinders Bay. All hands were saved, but the vessel became a complete wreck.[2] (See map on p. 20.)

The hull and cargo were put to auction in January 1879 at the same time as the two anchors and chain, recovered previously from the seabed at Lockeville.

NOTES

1. *Inquirer*, 31 July 1878.
2. Lionel Samson and Co., Auction Books, 1878, Acc. 1120A, p. 212, Battye Library.
3. *Inquirer*, 15 January 1879.
4. *Lloyds Shipping Register*, 1878, 1880.
5. Register of British Ships , Melbourne.

Annie Agnes

The cutter *Annie Agnes* was built at Fremantle in 1874 and operated in the coastal trade. In February 1878, she was stranded on the Moore River reef, but was refloated. She was in trouble again in October 1878, when she was stranded at Port Denison (Dongara) during a north-westerly gale.[1] This time, she stayed ashore for a considerable period. The *Inquirer* noted in 1885:

> The schooner *Annie Agnes* which has lain a wreck on the sand bank at Port Denison for 5 or 6 years has been put in thorough repair by the indefatigable diligence of the men who took the job in hand. The vessel was ready to launch a few days ago, but the late winds from the north-west have accumulated [sand] around her and will cause some extra labour before she is got afloat.[2]

Soon afterwards, the *Annie Agnes* was indeed refloated, and sailed with her new rig to Champion Bay, where she was reported to be 'as tight as a bottle'.[3] In November 1886, she was listed as a departure from Fremantle for Carnarvon. The vessel was finally wrecked at Lewis Island, 65 kilometres west of Cossack, in 1909.

The 33-ton *Annie Agnes* (Official Number 72466) originally had one deck, one mast and an oval stern.[4] The carvel-built vessel was 15.8 metres by 4.9 metres by 2.1 metres. For a time, she was owned by William Marmion.

NOTES

1. *Inquirer*, 9 October 1878.
2. *Inquirer*, 2 September 1885.
3. *Inquirer*, 14 October 1885.
4. Register of British Ships, Fremantle.

Canning

The flat *Canning* sank at her anchors in the South Bay at Fremantle during a gale early in November 1878.[1] The vessel had a cargo of jarrah aboard at the time. A flat is a barge or lighter. (See map on p. 242.)

NOTE

1. *Inquirer*, 13 November 1878.

Charlotte Padbury

The 636-ton barque *Charlotte Padbury*, a frequent visitor to Western Australian ports from London, stranded on the Point Moore Reef, at Champion Bay, while entering that port early in November 1878.[1] The vessel did not sustain major damage.

NOTE

1. *Inquirer*, 13 November 1878.

Firefly

The 9-ton cutter *Firefly* stranded on the beach at Fremantle during a gale early in November 1878. She was described as lying abreast of the Water Police Station. In 1874, the *Firefly*, under the command of one Sanderson, had sailed from Fremantle on a pearling cruise. The vessel does not appear on the Register of British Ships at Fremantle. (See map on p. 242.)

NOTE

1. *Inquirer*, 18 November 1878.

Macquarie

The barque *Macquarie* was built at Macquarie Harbour, Tasmania, in 1846.[1] By 1852, her rig had been changed to that of a schooner, and she then only carried two masts. In 1870, the vessel called through Fremantle on a voyage to Singapore.

Fremantle shipowner William Owston purchased the *Macquarie* in 1872 for the intercolonial trade. On several occasions, she was sent overseas: to Singapore with sandalwood, to Java, and to Colombo.

In August 1878, the *Macquarie* sailed from Colombo for Shark Bay with a general cargo. At Shark Bay the vessel was loaded with guano intended for trans-shipment to the larger *Chalgrove*, waiting at Champion Bay.[2] Once the cargo was fully loaded, the *Macquarie* commenced her voyage to Champion Bay. However, a gale developed and the vessel was wrecked on Levillain Shoal, at the northern end of Dirk Hartog Island, in November. Captain Spence and his crew all reached Dirk Hartog Island, but were then faced with an exhausting 3-day walk, with neither food nor water, towards the south end of the island, where another vessel was known to be loading. Eventually, they were picked up by the *Mary* in the South Passage.[3] (See map on p. 250.)

The wreck of the *Macquarie* has not been found. The Levillain Shoal, principally a sand shoal, is about 2 kilometres off Cape Levillain. The *Macquarie*

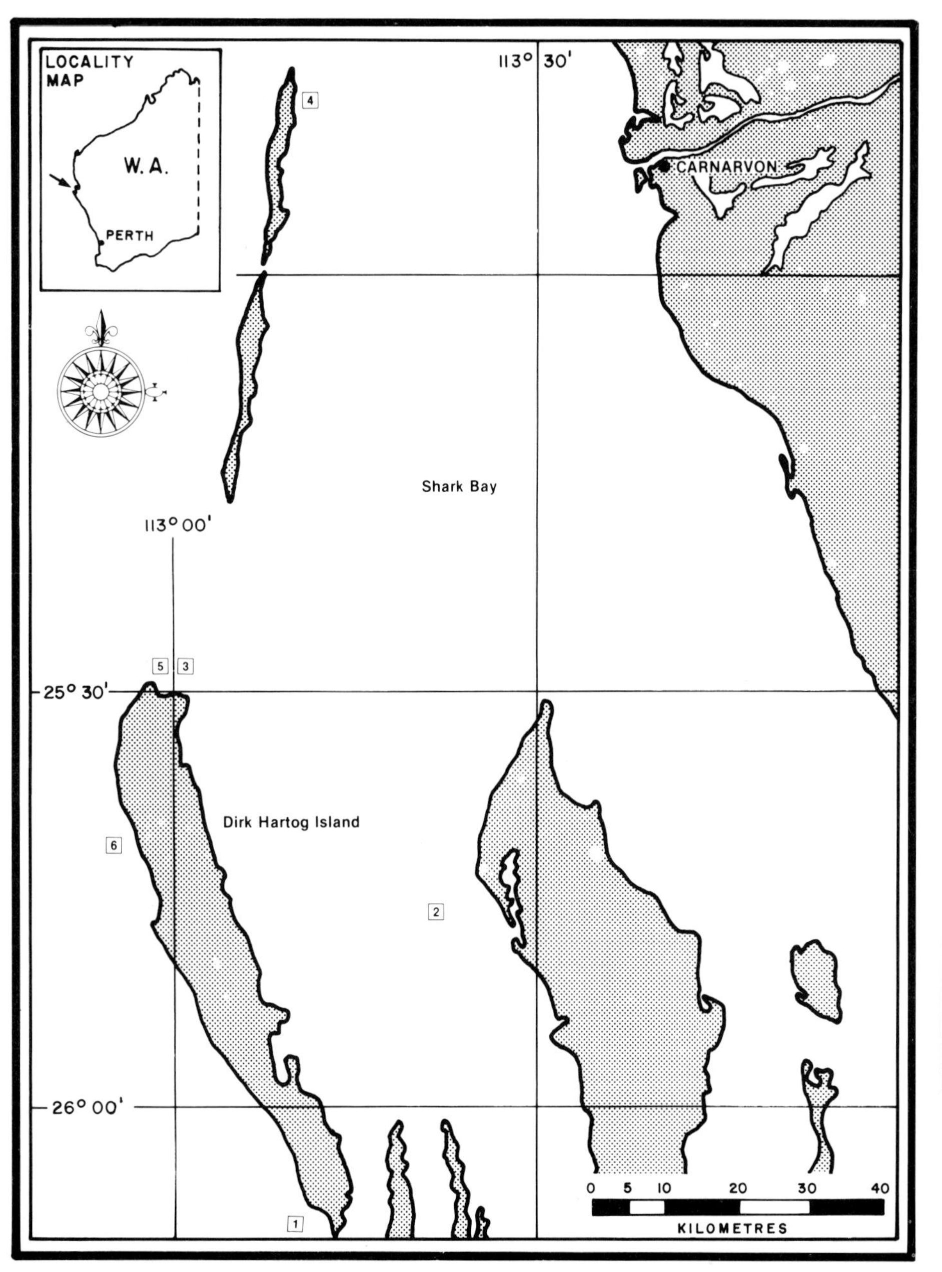

1 *Rover*
2 *Amelia*
3 *Perseverant*
4 *Paul Pry*
5 *Macquarie*
6 Unidentified Wreck Pre-1850

was 32 years old at the time of her loss, and had been ashore several times, so she might well have opened up and foundered on first bumping on the sand bottom.

The 125-ton *Macquarie* (Official Number 31967) was, at the time of her wreck, a schooner, with carvel build, square stern, one deck, two masts and a scroll head.[4] Her dimensions were 24.7 metres by 7 metres by 3.1 metres.

NOTE

1. Register of the *Macquarie*, CUS. 38, Tasmanian State Archives, Hobart.
2. *Inquirer*, 18 December 1878.
3. *Ibid*.
4. Register of British Ships, Fremantle.

SS Goolwa

Soon after departing Champion Bay on 13 December 1878, bound for Singapore with a cargo of horses in his schooner *Ariel*, Captain Hugo Leidicke saw an unusual vessel. It was a steamer, anchored about 10 kilometres off shore and making signals of distress. He shortened sail and made for the vessel, which turned out to be a steam hopper-barge from Glasgow owned by the South Australian Government. It was the 132-ton SS *Goolwa*, under the command of Captain Finch, with twenty hands on board.[1]

Captain Finch sent a boat across to the *Ariel* to explain that he was without coal or provisions. The steamer had left Ceylon on 1 November 1878, but instead of receiving the expected assistance of trade winds, the SS Goolwa ran into a series of southerly gales, and it was not until 27 November that the North West coast of Australia was sighted.[2] Falling short of coal, Captain Finch had been compelled to tear up the decks, bunks, bulkheads, ceilings, and all available timber for fuel. Then the barge had to fight her way down the coast against more contrary winds, progress being made even more difficult by the fact that, owing to the heavy consumption of coal, she was riding too light. Holes were cut in the sides of the vessel to fill the side compartments with water ballast, and in this way, she managed to reach the mouth of the Murchison River with no more than 1 day's supply of coal remaining.

A boat was sent ashore in search of wood for fuel, but it capsized in the surf, and the crew narrowly escaped from drowning. The coast abounded with sharks, and some of these were caught and used in the furnaces to raise steam. The SS *Goolwa* was anchored off Port Gregory, 65 kilometres north of Champion Bay, when the *Ariel* approached.

Captain Leidicke was not a man to let opportunity pass him by. He allowed Captain Finch to commission him to return the short distance to Geraldton to report the circumstances to the harbour authorities. Leidicke's fee for this service was £600![3]

The cutter *Moonlight* supplied the SS *Goolwa* with coal and provisions, and the steamer proceeded to Champion Bay for repairs. Three weeks later, she resumed her voyage, but had to take shelter from a gale in Bunker Bay. Putting to sea once more, she again encountered gales, which prevented her from getting past Cape D'Entrecasteaux and forced her to put back to Busselton for fresh supplies of coal, which she had been consuming at the rate of 11.5 tonnes a day. Leaving Busselton about the end of January, the SS *Goolwa* completed her long voyage without further mishap, arriving at Port Adelaide on 11 March.

The SS *Goolwa* (Official Number 79619) had been specially built in 1878 at Glasgow for the South Australian Government. The builders informed the Captain that she would steam at 18 kilometres per hour and was an admirable sea boat, but Captain Finch found that her speed under steam was a mere 7.2 kilometres per hour, and soon after leaving Aden, the crew had to work in a metre of water while getting out coal for the furnaces, because of a leak in the hull.

Two other barges were sent out to South Australia at the same time—the *Kadina* and the *Wallaroo*. The *Kadina*, like the SS *Goolwa*, ran short of fuel on the North West coast, became too light and made little headway against the wind. She called at Dirk Hartog Island for stone ballast—25 tonnes of which were conveyed at risk through heavy surf in the longboat—and to bury two men who had died of fever *en route*.

NOTES

1. *Inquirer*, 1 January 1879.
2. Notes on the Dredger *Wallaroo* and the Hopper Barges *Kadina* and SS *Goolwa*, imported by the South Australian Goverment in 1879, Acc. No. SAA/RN88, Libraries Board of South Australia.
3. Notes on the Dredges, SAA/ RN88, Libraries Board of South Australia.

Mariano

The condition of vessels employed in the guano industry was sometimes surprisingly poor, considering that these ships were sent to remote, unsurveyed ports for loading of bulk cargo. One example of such a ship's poor condition contributing to its loss is that of the *Mariano*.

The ship *Mariano*, bound from Melbourne to the Lacepedes in 1877, was forced to put in at Albany because of her leaky condition. The following year, she made another voyage to the Lacepedes. Captain William McDonnel loaded 750 tons of guano and set out for Melbourne on 23 November 1878.[1] On 5 December, the *Mariano* ran into heavy seas, with the wind from the south-south-west, and the ship began to leak at the rate of 30 centimetres per hour. Captain McDonnel shortened sail and kept both pumps constantly at work, but the water continued to gain, so he stood in for Tien Tsin, the closest port, to increase the crew's chances should the ship sink.

On arriving at Port Walcott, Captain McDonnel went ashore to make arrangements with his agents, F.M. McRae and Co., to bring her into Butcher's Inlet for unloading. The *Mariano* was then boarded by Captain Walcott, of the revenue cutter *Gertrude*, who undertook to pilot her in. Captain Walcott was the same mariner who, in 1876, conducted a search for shipwreck survivors after the loss of the *Stefano*. But on reaching the south side of Jarman Island on 13 December, the *Mariano* touched the ground and held there. It was at the time of the highest of the spring tides. Captain Walcott had sealed the fate of the *Mariano*.

She was securely moored, but little else could be done beyond waiting for the next spring tides. One hundred tons of cargo was taken out, but on the 22nd, she began to thump in such a manner that it was dangerous for men to stand beneath the spars. The officers and crew, in a subsequent letter, commented:

> . . . the captain, evidently fearing something, came on board and took his clothes and his chronometer on shore.[2]

The leak continued to gain on the pumps, and the crew by now were exhausted. At 10 o'clock the same night, the *Mariano* floated with 2.7 metres of water in her hold. First Officer H. French tried to bring the vessel into the channel, but she bumped and grounded close to Perseverance Rocks.

During the following day, the storm became a cyclone. One of the two ship's boats was smashed and made useless by the storm. The ensign was hoist with the Union Jack downwards, as a sign that the crew were desperately in need of assistance from shore. Captain McDonnel could not leave shore, because he was sick, but four Europeans and two Aborigines manned an open boat and rowed the 5 kilometres to the stricken vessel.[3] The crew returned to shore in the two boats, and the *Mariano* broke up completely that night. (See map on p. 118.)

The wreck was sold of £75. At an inquiry in Cossack, the master and officers were acquitted of all blame for the loss.

Little is known of the *Mariano* (Official Number 74670) herself. She was a 589-ton barque owned by Francis Beaver, of Melbourne.[4] *Registre Veritas* lists a 549-ton barque of that name, built at Boston in 1861 and rebuilt in 1868. That vessel measured 45.7 metres by 9.1 metres by 4.27 metres.[5]

NOTES

1. Cossack Station Report to Superintendent, 31 December 1878, Police Records, Acc. No. 129, Battye Library.
2. Officers and Crew of the *Mariano* to the Editor, the *Herald*, 15 February 1879.
3. *Ibid*.
4. *Mercantile Navy List*, 1879
5. *Registre Veritas*, 1878.

Runnymede

The Board of Trade Wreck Register states that a 640-ton Liverpool guano vessel (Official Number 29999) named *Runnymede* went ashore at Browse Island on 22 December 1878, her crew later being landed at Surabaya.[1] It was a victim of the same cyclone that destroyed the *Mariano* at Cossack. (See map on p. 152.)

The iron barque *Runnymede* was built in 1866 at Sunderland by Doxford, with the dimensions 51.7 metres by 8.9 metres by 5.7 metres.[2] The vessel, owned by J. M. Way and commanded by E. Way, was registered at Liverpool. Lloyds Register shows that she was built with thicker iron plates than the rules required.

Several vessels of the name *Runnymede* were in Australian waters during the 1870s, and this fact led the *Inquirer* to wrongly state that the whaler *Runnymede* was lost at Browse Island. That vessel, a 284-ton wooden barque (Official Number 32032), went ashore in Frenchman's Bay, King George Sound, in 1881. The *Runnymede* wrecked at Browse Island would have been loading guano. Yet another *Runnymede*, a 700-ton iron barque (Official Number 60940), plied Australian waters under the command of R. Hay.

Wreckage has been seen lying on Browse Island, in latitude 14° 07′ south, longitude 123° 33.6′ east. The reporters of that material were given a reward of $75 in 1975, and the site is protected under the Historic Shipwrecks Act. It is not yet possible to ascertain whether the site is that of the *Runnymede*.

NOTES

1. Board of Trade Wreck Register, National Maritime Museum, England.
2. *Lloyds Shipping Register*, 1877.

Sulina

The *Sulina*, under the command of Captain Ormond Thomas, was wrecked at Browse Island during a storm on 22 January 1879.[1] The vessel would have been loading guano at the time. No lives were lost, but little or no property was saved. The revenue cutter *Gertrude* took twenty-two of the crew to Cossack, where they were paid off. (See map on p. 118.)

The 1,142-ton wooden ship *Sulina* (Official Number 55022) was built at Nova Scotia in 1866, and was owned by G. B. Crow, of Liverpool.[2] Her dimensions were 54.8 metres by 11.6 metres by 7.3 metres.

NOTES

1. R.W. Vincent to Police Superintendent, Cossack, 14 March 1879, Police Records, Acc. No. 129, Battye Library.
2. *Lloyds Shipping Register*, 1879.

Manfred

The barque *Manfred* was loading guano for Messrs F. E. Beaver and Co. at the Lacepede Islands on 24 January 1879, when a gale blew up from the east and then veered to north-east. The vessel began dragging her anchors towards a reef, so Captain John Smith made the decision to run her into the 'gut', a passage between the western island and the inhabited island. In the process, the ship grounded on a sand-bar, and although the crew managed to get her off into deeper water, she was then driven onto rocks, where she became a total wreck. All of the crew made it safely ashore. (See map on p. 277.)

At the time of going ashore, the *Manfred* had 750 tonnes of guano aboard, and this was lost.[1] The wreck was sold at public auction for £205. Some of the crew were taken to Cossack on the cutter *Edward James*, while ten men remained with the guano loading settlement on the Lacepedes.

An inspection team examines the remains of the *Manfred* wreck at the Lacepedes.

The 587-ton *Manfred* (Official Number 47712) was built at Sunderland in 1864, and was owned by W. Davison, of North Shields.[2] The vessel had one deck, an elliptical stern, a figurehead of a woman, and the dimensions 46.4 metres by 9.3 metres by 5.9 metres.

At low-water, the iron deck supports, bollards, windlass and other components of the *Manfred* are obvious to visitors to the Lacepede Islands. The wreck was inspected by the Museum in 1978.[3] It lies in latitude 16° 51.4′ south, longitude 122° 07.7′ east, and is protected under the Historic Shipwrecks Act.

NOTES

1. Police Station, Cossack, to Police Superintendent, 14 March 1879, Police Records, Acc. No. 79.
2. *Lloyds Shipping Register*, 1879.
3. Scott Sledge, *Report of Wreck Inspection North Coast, 1978*, Western Australian Museum, Perth, 1980.

Industry

The cutter *Industry*, employed in pearling on the North West coast, was at North West Cape early in January 1879. On board were the master, Valentine Hester, two other Europeans and two Aborigines. The cutter started out for Port Walcott, and passed Mardie Creek on the morning of 20 January.[1]

The *Industry* continued on towards Port Walcott that afternoon. When the vessel arrived at Flying Foam Passage about 58 kilometres from Port Walcott, the barometer glass was falling ominously low, and Hester wisely decided against proceeding. The men secured the *Industry* and moored her, intending to continue on when the glass rose or the weather improved. But on the afternoon of the 24th, a heavy gale came on, and because the *Industry* was an old vessel, the five crew decided it would be safer to go ashore in their dinghy and camp overnight.

The gale lasted all night and the *Industry* was dashed to pieces in Boat Passage. On the 28th, her crew took the dinghy to Port Walcott, arriving the next day, and walked the 16 kilometres to Cossack. (See map on p. 118.)

The 12-ton *Industry* (Official Number 36539) was built at Perth in 1856 as a two-masted dandy.[2] Her dimensions were 11.7 metres by 3.5 metres by 1.3 metres. The vessel belonged to a Mr Hall.[3]

NOTES

1. R.W. Vincent (Roebourne) to Superintendent of Police, 11 February 1879, Police Records, Acc. No. 129, Battye Library.
2. Register of British Ships, Fremantle.
3. *Inquirer*, 19 March 1879.

Rosette

The 67-ton topsail schooner *Rosette* foundered at her anchors during the cyclone that struck the North West on 24–25 January 1879, and all of the nineteen men, women and children aboard her perished.

Since her launching at Fremantle in 1874, the *Rosette* had been a familiar sight at the North West ports. She left Fremantle on 7 January 1879 for Port Walcott, carrying on board thirteen passengers and six crew, and a general cargo for the pearlers. The explorer Alexander Forrest had wanted to take passage on her with his survey party, but he was not ready for departure when Captain Vincent wanted to sail. Vincent refused to wait, saying that the Exmouth Gulf pearlers were short of provisions and needed his cargo.[1]

Five days later, the *Rosette* rounded North West Cape and headed north-east towards Port Walcott. Captain Vincent knew the area well, but nevertheless allowed the *Rosette* to run aground on 16 January, on a sandbank at Thompson's Island, some 6 kilometres from the mainland and a few kilometres west of the mouth of the Fortescue River. When the pearling cutter *Industry* passed through that way on the 20th, Captain Vincent told the crew that the *Rosette* was quite safe on a soft sandbank, and that he expected to refloat her and continue his voyage that night. Some of the *Rosette*'s passengers considered transferring to the *Industry* for the remainder of the voyage to Port Walcott, but changed their minds, deciding that the swifter *Rosette* would arrive there first.

The *Industry* continued on towards Port Walcott and was wrecked in Flying Foam Passage. When the *Industry*'s crew arrived at Port Walcott and the news circulated as to the *Rosette*'s non-arrival, it was assumed that the *Rosette* must have drifted onto one of the islands west of Port Walcott during the gale, and several pearling vessels searched for her, unsuccessfully.

On 17 March, the schooner *Kate* left Port Walcott for Fremantle. She returned, the master, Matthew Kelly, reporting that he had found the schooner *Rosette* sunk at her anchors about 3 kilometres from Rosemary Island.[2] On 20 March, while standing out between Goodwyn Island and Rosemary Island, Kelly saw a mast poking above the water.[3] It had been cut away with an axe, but was entangled in rigging. A diver sent down to investigate made out the name *Rosette* on a hull lying perfectly upright in 9 metres. The foremast had been cut away, but the jib-boom was in perfect condition, as was the wheel. Her bow was facing the south-west. Kelly left a boat and two crewmen to watch the wreck, and returned to Port Walcott to claim salvage rights. (See map on p. 118.)

Captain Pemberton Walcott was instructed to take the revenue cutter *Gertrude* to inspect the wreck. He secured the services of four of the best breath-holding divers on the coast. Walcott later reported:

> . . . at 6.30 a.m. took my natives to the wreck and directed them to go down into the cabin. On reaching the surface they informed me that the companion way was blocked with bags and wreckage. Mr E. Chapman [who had led a boat crew to the rescue of the *Mariano* crew several months previously and was now at the site on Kelly's behalf to investigate salvage prospects] also went down and verified this. After several dives an entry was effected and the divers reported that all cabin fittings were gone and that the cabin was full of flour bags. As there was a very strong tide running and the natives seemed rather frightened I directed them to dive around the wreck and examine the hull. They reported the main hatch to be off and large portion of the cargo lying on the port side before the beam—tanks, flour, galvanised iron roofing, beer, pickles, wine, harness, being seen in quantity.[4]

Captain Walcott's crew found on Goodwyn Island shirts, hats, socks, towels and two kegs of matches. Then, on the western end of Enderby Island, they found eight stone-built graves, about half a metre high. On closer examination, the men decided that these were graves of whalers buried at least 20 years ago.[5] Lieutenant Helpman had seen three graves on Enderby Island in 1851, during his survey in the brig *Saucy Jack*.[6] It seems likely that these were some of the same graves seen by Walcott. When the whaler *Perseverant* was wrecked in 1841, the accident led to five men dying of scurvy on an island, but the records indicate that the island was Dirk Hartog Island, in Shark Bay. There are no records of whaling vessels having been wrecked near Enderby Island.

Considering the evidence, Captain Walcott concluded that the *Rosette* must have been anchored between Goodwyn and Rosemary Islands in the early part of the cyclone, and as the wind increased, the crew had discharged cargo and brought down the top-hamper. When the *Rosette* seemed in danger of being overpowered by the force of the wind, the crew had endeavoured to cut away the masts, but they had broken and gone by the board before the rigging could be completely detached. Some hours later, perhaps at the point of foundering, the crew had left the *Rosette* in a boat or on floating objects, with the intention of going ashore at Goodwyn Island. However, Walcott predicted the tide would have swept boats, bodies or wreckage on past Goodwyn Island towards Cape Preston, some 50 kilometres distant.

On 31 March, Frank Roy, of the schooner *Adur*, came into Roebourne with the news that he had found three human bones, along with other wreckage from the *Rosette*, at Cape Preston.[7]

Matthew Kelly hired the cutter *Robert James*, under the command of Edward Chapman, to return to the *Rosette* wreck site to conduct salvage, and his divers raised the Roebourne mails, along with a cash box that contained the diary of passenger Tom Ralston.

The *Rosette* (Official Number 61117) was built with one deck, two masts and an oval stern, and her dimensions were 21.5 metres by 5.5 metres by 2.4 metres.[8] She was owned by the Fremantle merchant William Dalgety Moore.

NOTES

1. A. Hicks, 'The Kimberleys Explored', *RWAHSJ*, 3 (1938), pp. 11–19.
2. R.W. Vincent (Roebourne, 7 April 1879) to Police Superintendent, Police Records, Acc. No. 129, Battye Library.
3. Sholl to Col. Sec., Roebourne, 25 March 1879, C.S.O. file 927/1879, fol. 2.
4. Pemberton Walcott to Col. Sec., 25 March, 1879, C.S.O. file 927/1879, fol. 2.
5. Walcott to Col. Sec., 25 March 1879, C.S.O. file 927/1879, fol. 2.
6. Exploration Diaries, Vol. 4, pp. 280–1, Battye Library.
7. Vincent to Police, 7 April 1878, Police Records.
8. Register of British Ships, Fremantle.

Unnamed Boat and Dinghies on North West Pearling Grounds

The January 1879 cyclone caused the destruction of other smaller craft besides the *Rosette* and *Industry*. The Roebourne correspondent in the *Inquirer* noted:

> On the pearling ground one of Mr McRae's boats was wrecked and another damaged. A good many dinghies were lost, but no lives. All the shells were saved.[1]

NOTE

1. *Inquirer*, 19 March 1879.

SS Start

The screw steamship *Start* was one of Western Australia's more ambitious early ventures into steam. The vessel was built at Fremantle in 1876 and went as far afield as Melbourne in 1878, but it did not last long. The *Inquirer*, in March 1879, reported:

> This unfortunate little steamer of whose probable fate so much has been conjectured, there is now but little doubt has foundered. Mr Scott, the Asst. Telegraph Operator at Israelite Bay, while traversing the coast here on Friday last, came across an oar branded '*Start*' and a vessel's stern post branded [with] the same name. This is undoubtedly a portion of the vessel, which was last spoken by the *Cleopatra* on her way to this colony from Melbourne. The Pollock Reef is between 50 and 60 miles off Israelite Bay, where it is probable that the unfortunate craft met her sad fate and Captain Allen and his crew came to their untimely doom.[1]

The *SS Start* (Official Number 75289) was carvel built, of wood, with an oval counter stern.[2] She had one deck with two masts, and was rigged as a fore-and-aft schooner. The vessel was 23.6 metres by 5.4 metres by 2.2 metres, and measured 66 gross tons. She had a 20-horsepower steam engine and was screw propelled. The owners of the *Start* were Fremantle merchants William Dalgety Moore and William Marmion, and Fremantle shipbuilder James Storey. (See map on p. 198.)

NOTES

1. *Inquirer*, 19 March 1879.
2. Register of British Ships, Fremantle.

Cock of the North

It would appear that another vessel was wrecked at Point Cloates early in 1879. The *Inquirer* reported:

> . . . there has been another wreck in the neighborhood of the North West Cape, at a place called Point Cloates. The *Planet*, on her last trip from Fremantle put in there, the Captain having described some wreckage on the beach. Upon landing he found the beach strewed with coconuts, Manilla rope, and rigging which indicated that the wrecked vessel was a brig. No traces of survivors were discovered. There is no doubt many disastrous wrecks occur on this coast and are never heard of . . .[1]

A later article suggested that the wreckage belonged to a brig named the *Cock of the North*, but no particulars were given as to her port of departure or destination.[2] (See map on p. 30.)

Lloyds Shipping Register lists a 235-ton three-masted schooner (Official Number 76564) of that name built at Kingston in 1877.[3] However, it was afloat in 1882, when it visited Port Adelaide.[4] That vessel was felted and yellow-metalled in the same year, and had the dimensions 35.4 metres by 7.3 metres by 4.1 metres. Another vessel of the same name, described as a smack, was later employed by Charles Broadhurst at Shark Bay.

NOTES

1. *Inquirer*, 7 May 1879.
2. *Inquirer*, 4 June 1879.
3. *Lloyds Shipping Register*, 1879–80.
4. Shipping Register, Port Adelaide, Acc. No. 436, South Australian Archives.

Mary

The schooner *Mary* was launched, amidst the cheers of a large assemblage of spectators, at Fremantle in December 1868. The vessel was used in the coasting trade, carrying pearl shell and wool from North West ports to Fremantle.

In May 1879, the *Mary* was at Binningup, to the north of Bunbury, loading timber.[1] Bad weather prompted the *Mary*'s master, John Waldron, to seek shelter at Lockeville, tying up alongside the Lockeville timber jetty on 4 June (a timber mill operated nearby at Wonnerup). When it became squally, he hauled the vessel off to the West Australian Timber Company's mooring, some 180 metres from the jetty. By 2 a.m. on 5 June, a heavy gale was sending large quantities of water right over the *Mary*, and she dragged, striking one of the jetty piles and going broadside onto the jetty. This made it impossible to bring the vessel around to the lee side of the jetty. (See map on p. 204.)

Waldron paid out his cable to let the *Mary* back gradually onto the beach. Soon after she commenced striking the bottom, the masts broke, so the crew scuttled the vessel, which steadied her.

When the storm subsided, the *Mary* was a wreck, and the position of her hull prevented repairs being done to the jetty. The manager of the Timber Company called upon Harbour Master George Forsyth to order the removal of the wreck, but the Government's legal advice was that the legislation did not necessarily apply to private jetties.[2] The Timber Company solved the problem by purchasing the wreck.

The remains of the *Mary* (consisting of the keel and sternpost) can still be seen lying near the jetty piles. The jarrah rudder was pulled from the water by Busselton fishermen in the 1960s.

The 48-ton two-masted schooner *Mary* (Official Number 61087) was built at Fremantle in 1868. She had one deck, an elliptical stern, and dimensions of 20 metres by 5.2 metres by 2.3 metres.[3] In 1879, she was owned by William Pearse and partners, of Fremantle.

NOTES

1. Inquiry into casualty to *Mary*, 5 June 1879, C.S.O. file 1057/1879, fol. 4.
2. George Forsyth to Col. Sec., 7 June 1879, C.S.O. file 1057/1879, fol. 1.
3. Register of British Ships, Fremantle.

Belle of Bunbury

The 42-ton coastal schooner *Belle of Bunbury* was stranded at Bunbury in August 1879. The *Inquirer* reported:

> On Tuesday last the *Belle of Bunbury* loaded with timber was struck in a heavy squall, which came on quite unexpectedly, and broke her anchor. She came down on the jetty, causing considerable damage to her rigging, bulwarks etc as well as to the jetty.
>
> Had it not been for the great assistance rendered by the Water Police, and a number of residents in the town who all worked with a will, the damage would have been much greater.[1]

The vessel was repaired and put back into service, and was finally wrecked in 1886.

NOTE

1. *Inquirer*, 13 August 1879.

Knowsley Hall

Loney, in his *Australian Shipwrecks*, notes the possibility that the 1,774-ton iron ship *Knowsley Hall*, which left London in June 1879 with fifty-five passengers for New Zealand, was wrecked off the south coast of Western Australia.[1] In August 1881, a portion of her bulwarks, bearing the painted letters 'KNOW' and 'S', was found at Point D'Entrecasteaux. It is, of course, possible that the wreckage, like the figurehead of the *Blue Jacket*, drifted thousands of kilometres before beaching.

NOTE

1. Jack Loney, *Australian Shipwrecks, Volume 3: 1871 to 1900* (Geelong, 1982), p. 108

Ione

The 25-ton schooner *Ione* was stranded at Bunbury during a storm on 2 October 1879. The vessel broke an anchor and drifted onto the jetty, through which she ultimately forced a passage.[1] She then continued on towards the beach, finally stopping near the wreck of the *Midas*. The stern of the *Ione* was broken away, but she made no water, and was refloated 4 days later. The vessel was reported back at Fremantle on 15 October.[2]

NOTES

1. Inquiry respecting *Ione*, Bunbury, 9 October 1879, C.S.O. 456/1879, fol. 25.
2. Register of Arrivals.

Sarah Burnyeat

The 277-ton barque *Sarah Burnyeat*, loaded with jarrah piles for Capetown, was driven ashore at Lockeville during a heavy gale on 3 October 1879.[1] The captain, second mate and three of the crew were ashore at the time. She was condemned as a wreck, but 5 weeks later, the *Sarah Burnyeat* had been refloated, and in January 1880, the dismasted vessel was towed to Fremantle for repairs. After lengthy repairs she set sail for Port Adelaide in October 1881, and sprang a leak.[2] In 1882, she sailed with coals to Albany on her final voyage.[3] Henceforth, she was used as a coal hulk.

NOTES

1. Police Station, Vasse, to Police Superintendent, 4 October 1879, Police Records, Acc. No. 129, Battye Library.
2. *Inquirer*, 7 December 1881.
3. *West Australian*, 5 May 1882.

Ben Ledi

The iron ship *Ben Ledi* left Sydney on 8 November 1879, in ballast, bound for Calcutta, with a crew of twenty-three. Captain John Boyd headed east across the Great Australian Bight and around Cape Leeuwin. At noon on 16 December, the *Ben Ledi* was about 105 kilometres south of the Abrolhos Islands.[1] At a few minutes before 7 p.m., as the sun was setting, Captain Boyd went aloft to the main topgallant mast-head, and with his telescope, saw the mainland at a distance he judged to be 68 kilometres, which would, he thought, have put him clear of the Abrolhos on his intended course. From 7 p.m., the ship was sailed north-west by west until striking, at 11 p.m., on the south end of Pelsart Island. The *Ben Ledi* was travelling at 15 kilometres per hour when she struck, and she remained fast.

Twenty-four hours later, the ship had made no water, but she lay on a shelf of rock and there was no possibility of refloating her. Captain Boyd and five of the crew arrived at Geraldton in the ship's lifeboat early on the 19th. A court of inquiry was told of currents in the area and found no reason to charge Captain Boyd. The ship was reported to have been insured for £13,000, but the wreck was sold at auction for £80.[2]

The 1,107-ton *Ben Ledi* (Official Number 60339) was built in 1868 at Glasgow by Barclay, Curle and Co.[3] Her dimensions were 66.4 metres by 10.5 metres by 6.4 metres, and she had two decks.

When inspected by the Museum, substantial areas of iron plating and frames could be distinguished on the seabed. The vessel appears to have struck in an upright position, and the sides have subsequently fallen out. The site, lying in latitude 28° 55.5′ south, longitude 113° 59.6′ east, is protected under the Historic Shipwrecks Act. (See map on p. 44.)

NOTES

1. Inquiry respecting *Ben Ledi*, Geraldton, 8 January 1880, C.S.O. 1309/1880, fol. 46.
2. *West Australian*, 3 February 1880.
3. Lloyds Survey Register.

Museum archaeologist Mike McCarthy surveys the *Ben Ledi* wreck.

Adalia

Goldsmith, in his *Treasure Lies Buried Here*, refers to a schooner named *Adalia* being wrecked in a cyclone of 9 January 1880 near Yammadery Creek, with the loss of one life.[1] A tidal wave 7.6 metres over high-water mark is said to have been recorded. It seems that Goldsmith was mistaken, because an article in the *Herald* of February 1881 mentions that a vessel named *Adela* had been capsized by a 9.1 metre high tidal wave resulting from a cyclone that commenced on 7 January 1881.[2]

A 10-ton cutter named *Idalia*, owned by C. Presswell, was operating in the pearling industry in the 1880s, so, assuming that all three vessels are the same, it is likely that the vessel capsized by the tidal wave was later refloated.[3]

NOTES

1. Frank Goldsmith, *Treasure Lies Buried Here* (Perth, 1946), p. 201.
2. *Herald*, 12 February 1881, p. 3f.
3. Report by the Inspector of Pearl Shell Fisheries, 1887–8. *Western Australian Blue Book*, 1889, p. 531.

Aldergrove

Grolier's *Australian Encyclopedia* records that the 1,271-ton iron barque *Aldergrove* stranded off Point Malcolm, at the head of the Great Australian Bight, on 3 May 1880.[1] The vessel was refloated and survived until 1905, when she was lost off Peru.

NOTE

1. Grolier's *Australian Encyclopedia*, Volume 10 (Grolier Society, Sydney, 1965 edition), p. 5.

PS Endeavour

The first steam vessel to be used exclusively for goods traffic on the Swan River was introduced in February 1877, and the *Western Australian Times* gave a description of the newcomer:

> A new river steamer named the *Endeavour* has been built by Mr Robert Wrightson. Its length is 76 feet with about 5 feet of hold. The boiler, tested in the sister colonies is capable of working a 24 horsepower engine. The power of the one she is to carry will be 20 horsepower. She is the lightest draft vessel for her size in the colony. Without boiler and engine she draws 10 inches, when all gear is on board her draft will be 17 inches. She is strongly built, and is the first steam vessel to be used exclusively for goods traffic on the river. She is alike creditable to her owner and builder. Her presence on the river will give an impetus to that trade and occasion a competition that the old steamboat proprietors will not take to very kindly.[1]

The *Endeavour* sank in 1880. She had been moored at North Fremantle on 3 June, awaiting the morning tide to pass over the Bar, but a heavy north-west gale during the night drove her against the railway bridge, and she foundered.[2] She had on board a large quantity of cargo intended for shipment by the *Clarence Packet* to Port Walcott. Soon after the vessel went down, it was reported that she had been 'outrageously plundered of a number of cases of sardines', and that 'the police have under their eye two or three very suspicious characters supposed to be connected with the matter'.[3] A good deal of the cargo was legally recovered from the wreck, but a week after the accident, the vessel had not been raised. The Board of Trade Wreck Register indicates that the vessel was saved, but the Register of British Ships states that she was untraced after 1879.[4] (See map on p. 242.)

The 31-ton paddle steamer *Endeavour* (Official Number 75295) was carvel built, with a sharp stern, one deck and a fore-and-aft-rigged mast. She was originally owned by York merchant John Monger, who sold her to Godfrey Knight.[5]

NOTES

1. *Western Australian Times*, 6 February 1877.
2. *Inquirer*, 9 June 1880.
3. *Ibid.*
4. Board of Trade Wreck Register, 1880. See also Register of British Ships, Fremantle.
5. Register of British Ships, Fremantle.

Planet

The 65-ton topsail schooner *Planet* was driven ashore at Geraldton on 4 June 1880 while *en route* from Cossack to Fremantle.[1] She was soon refloated, and arrived at Fremantle on 17 June.

NOTE

1. *Inquirer*, 9 June 1880.

Batoe Bassi

A telegram announced the stranding of the barque *Batoe Bassi* on 5 June 1880 near Inshore Island, to the east of Esperance.[1] The *Batoe Bassi* had been bound from Tjilitjap, in Java, for Melbourne with a cargo of 300 tons of sugar, 3,000 coconuts, rice and spirits. The vessel sprang a leak at sea, but the Malay crew were too sick to pump the vessel out. Two of the twenty-four crew had died during the voyage and most of the rest were laid up with what was described as rheumatism.[2]

Captain B. H. Buir had no alternative but to beach her. He ran the vessel ashore, and south-easterly winds filled her with water, destroying the cargo. The crew, including three women, were all safely landed. The mate, A. Brinkmann, sought assistance in Albany to refloat the *Batoe Bassi*, but he was reluctant to commit himself to pay the £100 charter fee for the 50-ton *Agnes*, so the Government ordered the telegraph assistant at Esperance Bay to go to the wreck with a field telegraph instrument, to place the captain in communication with his consignors in Melbourne.

The vessel was soon a complete wreck, its back broken. The hull was sold at auction at Albany and realized a mere £3, while the vessel's gear was sold at a nominal value. (See map on p. 198.)

The 293-ton *Batoe Bassi* was built of timber at Drammen in 1864.[3] Her dimensions were 35 metres by 7.5 metres by 4.12 metres.

The wreck was rediscovered in 1969 and reported to the Museum by two groups of divers, and several items, including coconuts, were raised. Part of the timber hull is still recognizable on the seabed, in latitude 33° 54.6′ south, longitide 122° 50.2′ east. The site is protected under the Historic Shipwrecks Act.

NOTES

1. *Inquirer*, 9 June 1880.
2. *West Australian*, 4 June 1880.
3. *Registre Veritas*, 1880.

Mayflower

The brig *Mayflower* was built by W. Pickersgill, of Sunderland, in 1867 for G. Lawson. In about 1873, the vessel was sold to Captain Peter Dickson, of Port Adelaide. Dickson went to London with his wife and two children to collect the vessel, and at Antwerp, he loaded a cargo of dynamite for Australia at a favourable rate.[1] When the vessel arrived at Williamstown, the port authorities were aghast at the nature of the cargo and ordered that it be discharged into lighters in Port Phillip Bay. Eventually, the *Mayflower* sailed into the Port River at Adelaide, and the cargo was discharged into drays in the normal way.

Captain Dickson passed command on to a Captain Richards, who successfully traded between Mauritius, the Cape, New Zealand and Australian ports for the next 6 years, before in turn handing over to Captain William Walker, in May 1880. Walker left Table Mountain for the stated destination of Guam in the central Pacific, but he lost the *Mayflower* at Augusta in July 1880.[2]

Walker described in the log book the circumstances after he anchored in Flinders Bay, an area exposed to south-easterly winds:

> 14 July Flinders Bay 12 p.m. [midnight] Moderate breeze at SSE and squally weather with showers of rain and heavy ground swell rolling into the Bay.

The *Mayflower*

> 1.15 a.m. The vessel touched ground. Called all hands and set jib, T. Staysail and main topmast staysail and tried to cant the vessel to the eastward to get port anchor down. The vessel would only cant about 4 points. About 3 a.m. let go port anchor, the vessel kept clear of the ground for the time. At daylight sent on shore for a boat to carry out stream anchor, having first run out kedge and line with our own boat. About 7 a.m. the vessel began to touch the reef again. Run out stream anchor and 3 inch coir hawser. Hove taut and hove ahead on both anchors, wind freshening up at South with strong swell rolling into the Bay. About 8 a.m. heaving in chain the pawls of the windlass carried away. Secured the chains to fore and main masts. The windlass bitts having carried away and rudder unshipped at 9 a.m. port chain cable carried away, the vessel dropping astern and bumping heavily on the reef, 12 inches water in the well.
>
> 10 a.m. 2 feet of water in the well and the vessel striking the ground heavily. All hands at the pumps for an hour but found the water gaining and the sternpost started.
>
> 12 a.m. [noon] Every appearance of the vessel becoming a total wreck. Hoisted out another boat and prepared to leave the vessel. Sent on shore the personal effects of the crew.
>
> 2 p.m. The water in the hold over the beams, the vessel striking heavily and the masts in danger of going by the board. Left the brig in her own boats and so landed at Port Augusta.[3]

The *Mayflower* did indeed become a total wreck. A court of inquiry found Captain Walker guilty of default in anchoring the vessel too near the shore, and suspended his certificate of competency for 6 months.[4] The wreck was sold at auction at the Vasse Hotel in Busselton.

The 277-ton *Mayflower* (Official Number 56115) was carvel built, with one deck and a round stern.[5] Her dimensions were 33.5 metres by 8.1 metres by 4.9 metres.

In modern times, the wreck was found by Dr John Williams, of Augusta, at Deere Reef to the north of Barrack Point, in latitude 34° 20′ south, longitude 115° 10.5′ east. The site is protected under the Historic Shipwrecks Act. A reward of $200 was paid to the finder. (See map on p. 20.)

NOTES

1. D. R. Dickson, Notes on the *Mayflower*.
2. Guam was not a real destination; it was used when ships' masters or owners did not want to divulge their real destination, or if a vessel went 'seeking' cargo in South-east or East Asia.
3. Extract from *Mayflower* Log, Busselton Courthouse Records, Acc. No. 594, Battye Library.
4. *Inquirer*, 18 August 1880.
5. Register of British Ships, Port Adelaide.

Moorburg

The 227-ton German schooner *Moorburg* was driven ashore in Owens Anchorage during a gale in July 1880, and was firmly fixed in the sand.[1] After unloading, she was refloated and sailed for Hong Kong.

The *Moorburg* captured the attention of the local Perth press on several other occasions. Prior to her arrival at Fremantle, the vessel called at Albany. There, Governor Robinson presented to Mrs Maria Boldt, wife of the captain, the New South Wales Relief Society's Silver Medal in recognition of her bravery. The *Herald* outlined the circumstances:

> The *Moorburg*, manned by a mate, four seamen and a Chinese cook sailed from China to Melbourne in July last, with a cargo of tea. During the voyage intensely hot weather was experienced. All the livestock on board died from the heat, and the four seamen also died one after the other. The captain and the mate were likewise ill, and only just able to crawl about.
>
> Mrs Boldt, although a thin, delicate looking lady, proved equal to the dreadfully trying circumstances. She nursed the sick and cheered them, and when a gale which the schooner encountered after passing the Solomon Islands, came upon her, she took her position at the helm, and single handed, steered the vessel through the storm. Subsequently, until the vessel arrived in port, she took her regular turn at the wheel and did a man's part in navigating the vessel.[2]

In those male chauvinistic, times it was considered extraordinary that a mere woman should be able to cope under such duress. Mrs Boldt was given a locket set with diamonds, and a necklet of gold.

Just two months after leaving Fremantle, the *Moorburg* was again in the news. The Batavia *Handelsblad* reported:

> On the 7th September there stranded on Belvidere Reef, near Gaspar Island, the German schooner *Moorburg*, laden with sandalwood and bound from Fremantle to Hong Kong. The crew have been brought to Anjer by the American ship *Abbel Abbot*.[3]

Perhaps Mrs Boldt should have retained her earlier control of the helm, or persuaded her husband to take a shore job after assisting with his navigation.

NOTES

1. *Herald*, 26 June 1880.
2. *Ibid*.
3. Batavia *Handelsblad*, 15 September 1880, reprinted in *Inquirer*, 3 November 1880.

Albert

The 23-ton cutter *Albert* was built in 1861 on the Swan River at Perth and initially used for local trading. On 1 August 1867, the vessel departed Fremantle for Champion Bay carrying 225 kilograms of gunpowder.[1] From Geraldton, the *Albert* had been chartered to proceed to Roebourne, but while in port at Geraldton, the government arranged for her to search the Abrolhos for shipwrecks. Two vessels, the *Brothers* and the *Emily*, were missing.

The *Albert* left for the Abrolhos on 8 August, and the next morning, the crew sighted the Middle Group, but heavy squalls hindered their progress. On the morning of the 11th, the vessel struck a reef a kilometre from a small island in the Northern Group called Goss' Monument, which bore east by north.

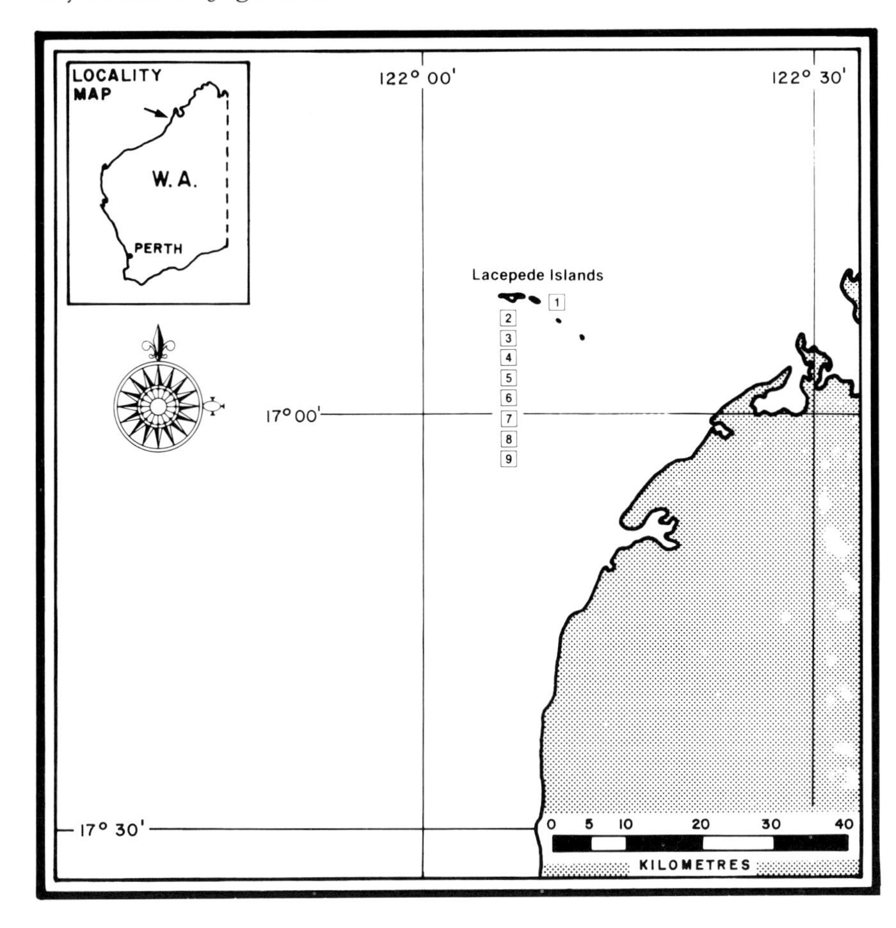

1 *Albert*
2 *MANFRED*
3 *Aboyne*
4 *Isabellas*
5 *Helen Malcolm*
6 *Albert Victor*
7 *Start*
8 Three Lighters 1877
9 Ship's Boat 1877

Goss' Monument, presumably so named after Captain Goss, who camped there when the *Favourite* was stranded, was later renamed Beacon Island. After lowering the sails and launching the dinghy, the crew took the provisions and water ashore, and began discharging ballast from the *Albert*. From the 13th to the 19th, the crew tried to get the vessel afloat, without success. Then they made a mast and sail for the dinghy, and three men set out for Geraldton.[2]

Arrangements were made at Geraldton for the *Swan* to return to the Abrolhos to refloat the *Albert*, which lay in a sheltered position on soft coral. The cutter *Victoria* also lent a hand in refloating the *Albert*, before returning to Fremantle on 16 September.[3] The *Albert* then resumed her coastal trading. In 1875, the vessel was involved in the pearling industry at Exmouth Gulf when a fracas broke out on board, resulting in several deaths.

The *Albert* was eventually totally wrecked at the Lacepedes on 31 January 1880, while engaged in the pearling industry. Discussing the loss in April 1881, the Collector of Customs, J. W. Clifton, lamented:

> It is quite impossible for me to obtain the necessary information to fill in 'Wreck Forms' for casualties happening to Pearlers on the N. West Coast when no Court of Inquiry is held and the bare fact comes to my knowledge many months and sometimes over a year after the event.[4]

The *Albert* (Official Number 36544) was carvel built, with a square stern, and measured 13.8 metres by 4.4 metres by 1.7 metres.[5] Her last shareholders were F. McRae, G. Fauntleroy and F. Pearce of Roebourne.

NOTES

1. *Inquirer*, 7 August 1867.
2. Journal of P.C. Thomson, 23 August 1867, Police Files, Acc. No. 129, Battye Library.
3. *Inquirer*, 9 October 1867.
4. J.W. Clifton to Col. Sec., 22 April 1881, C.S.O. 1309/1881, fol. 88.
5. Register of British Ships, Fremantle.

Citizen of London

The 53-ton schooner *Citizen of London* was built of jarrah by former London draper George Payne on his property, more recently known as Lexton Park, a few kilometres from Capel.[1] Once he had assembled the craft, Payne then took it apart plank by plank and had it carted 14 kilometres to the Vasse property of James Forrest, where it was reassembled. George Payne's wheelwright son Arthur succeeded his father as master of the *Citizen of London* and traded on the coast for several years, but he held no certificate and the vessel was not insured.

On 20 August 1880, the *Citizen of London* was alongside the Bunbury jetty, loading with sandalwood for Fremantle. There was about 24 tonnes on board, but loading was made difficult by a nasty swell. George Payne was supervising the work:

> ... had taken in one truck load which I with my three men was down below stowing, when the vessel gave a tremendous surge against the jetty—I went on deck and found that the rope fender had crushed in three planks on the port side. The chief part of the injury was below the water line. We hove off from the jetty and from the dinghy we tried to stop the leakage with blankets. The blankets were sucked in under the skin, by the rush of water through. I then tried to place the flying jib against the damaged part but that also was sucked in. We then got sail on her and made for the north beach, about two miles distant from the jetty. We then ran her direct on shore. At that time there was 5 feet of water in her hold. We made things secure and got lines from both mastheads to the shore to prevent her rolling.[2]

There was a large surf on the beach, but Payne and his assistants discharged the cargo and other gear the following day, and subsequently shifted the vessel 5 or 6 metres up the beach for the purpose of repairing her. The wreck of the *Annie M. Young* lay nearby. (See map on p. 204.)

A court of inquiry considered that Payne had shown a lack of judgement in not warping the *Citizen of London* up alongside the jetty until she grounded, and had thereby exposed the vessel to becoming a wreck on the beach in the event of gales.

In May 1882, it was reported that the *Citizen of London* had had very severe damage repaired, and been shifted preparatory to being refloated.[3] But the *Citizen of London* stayed on the beach. Explosives were used to break up the hull and her registry was closed in 1884.[4]

The fore-and-aft schooner *Citizen of London* (Official Number 75303) was built in 1878 and was owned by the father and son who sailed her.[5] She was carvel built, with one deck, two masts and a counter stern. The *Citizen of London*'s dimensions were 17.4 metres by 5.7 metres by 2.6 metres.

NOTES

1. *The Countryman*, 27 June 1961.
2. Inquiry respecting *Citizen of London*, 3 September 1880, C.S.O. 1309/1880, fol. 44.
3. *West Australian*, 5 May 1880.
4. Register of British Ships, Fremantle.
5. Register of British Ships, Fremantle.

Star

Veteran Fremantle boatbuilder Thomas Mews launched Messrs J. and W. Bateman's new schooner, named the *Star*, from his Arthur Head slipway in March 1876. It was schooner rigged and built for fast sailing.[1]

Batemans were initially unsure as to where to employ the *Star*. They first sent her to Batavia with a cargo of jarrah, but during the second half of 1877, the vessel was whaling, initially at the Rosemary Islands. That area had been fished by overseas whalers before the first settlement of Western Australia. In 1870, Messrs Pearse and Marmion were the first Western Australian whalers to exploit these grounds. John Bateman began operations there in 1872, using Malus Island as a base. The family had earlier been bay whaling from Port Gregory. The *Star*'s voyage in 1877 was successful, and she arrived home at Fremantle with 147 casks of oil.[2]

Mid-way through 1880, the *Star* was fitted out for a short whaling cruise to the south of Fremantle. Whales had been seen at Geographe Bay, so on 28 September, she left port under the command of Captain John Sheppard to commence the hunt. Captain Sheppard had no mate aboard, and the ship's crew were all Malays. The vessel carried two whale boat crews and a spare whaling hand, but they had nothing to do with the working of the ship. John Bateman senior was on board as one of the headsmen.[3]. Eight whales were sighted and closed upon, but they eluded capture, so at 3 p.m. on 19 October, the course was set for Fremantle.

John Bateman, an old hand in these waters, gave Captain Sheppard some advice on the course, warning him that there was a strong current set in towards shore and telling him not to go too far to the eastward. Captain Sheppard steered north-north-east for the Rottnest light, and the *Star* was pushed along at 11 kilometres per hour by a strong south-westerly wind. The ship's crew had been divided into two watches, two men being in Sheppard's watch and one in the other. At 1 a.m., Cape Bouvard was seen, bearing east, 11 or 12 kilometres distant. The sounding lead was lying on the hatchway close at hand, but it was not used.[4]

The Captain had a glass of brandy below in the cabin with Bateman, and went on deck again at 1 a.m. He altered the course two points to the east, believing that this would bring the vessel to the north end of Garden Island. A Malay was at the wheel and another was on the lookout. It was a moonlit night. Sheppard, thinking all was well, lay down on the deck and went to sleep.

At 3.15 a.m., the Malay helmsman saw breakers on the starboard bow. He shouted to the Captain, 'rocks on the starboard', and put the helm up. The schooner had begun to pay off towards the north when the Captain sprang up, shoved the Malay away from the helm, seized it, and in the confusion of the moment jammed it down. The schooner at once swung round heavily onto the reef, striking it so violently that the men were pitched out of their berths.[5]

The *Star* was on the weather side of the Murray Reef, but she ground against the rocks for nearly an hour before the sturdy hull was pierced and the vessel

The *Star* in heavy seas.

suddenly sank in about 3.7 metres, her deck being just awash. The men saved their traps and all the sails, and set out in the boats for Becher Point and thence to Fremantle.

Soon after, it was reported that some of the whaling gear, including the try-pots and the firing apparatus, was lost.[6] However, Bateman hoped to save all the running gear, anchors and chains. The vessel was uninsured. Batemans had lost at least four other vessels up to that date, including the *Favourite*, the *Flying Foam*, the *Twinkling Star* and the *Bungaree*, this last vessel having been lost within several kilometres of the *Star*.

A preliminary inquiry into the loss of the *Star* heard evidence from Bateman that, in his opinion, the schooner had been lost because Captain Sheppard had altered the course. The situation looked bad for Captain Sheppard. Just prior to the second hearing, the *Inquirer*, in a biased article, noted that Sheppard was generally liked among the waterside folk, and argued that the set of the currents onto the Murray Reef was so powerful as to baffle entirely the calculations of any mariner.[7] The court was clearly not influenced by the press, because it found Captain Sheppard guilty of four charges: he had improperly altered course by two points on the mere assumption that he had sailed 96 kilometres from Eagle Bay; he had shown neglect in not having a proper lookout kept; he had shown neglect in not having taken cross-bearings; and he had shown neglect in not taking soundings when he was in doubt as to the position of the vessel. His certificate was suspended for 18 months and he was made to pay costs.

The 70-ton two-masted fore-and-aft schooner *Star* (Official Number 72482) had one deck and an oval counter stern, and the dimensions 24.1 metres by 5.3 metres by 2.3 metres.[8] The site of the *Star* wreck, lying in latitude 32° 22.2′ south, longitude 115° 41.2′ east, was found in 1973 by divers Graham Anderton, Bill Evans, Ross Morgan and D. Grove. They were given a reward of $50 dollars and contracted to do a preliminary survey of the site, which is protected under the Historic Shipwrecks Act. (See map on p. 242.)

NOTES

1. *Western Australian Times*, 10 March 1876.
2. *Western Australian Times*, 23 November 1877.
3. *Herald*, 23 October 1880.
4. 'Inquiry respecting *Star*', *Herald*, 12 November 1880.
5. *Herald*, 23 October 1880.
6. *Ibid*.
7. *Inquirer*, 17 November 1880.
8. Register of British Ships, Fremantle.

Diver Graham Anderton works on the *Star* wreck during a Museum excavation.

Bibliography

MANUSCRIPT SOURCES

Official

Bevaqua, B. Archaeological Survey of Sites Relating to the *Batavia* Shipwreck, May 1974. Western Australian Museum.
Board of Trade Transcripts. Acc. No. 108/67, Public Record Office, Kew.
Board of Trade Wreck Register, various dates. National Maritime Museum, England.
Busselton Courthouse Records. Acc. No. 594, Battye Library.
Certificate of British Registry, *Sara*. Acc. No. HS100, Battye Library.
Colonial Secretary's Office Correspondence, 1829–1890, Western Australia. Acc. Nos 36 and 527, Battye Library.
Exploration Diaries. Acc. No. PR5441, Battye Library.
Foreign Office Correspondence. Public Record Office, Kew.
Fremantle Harbour Master's Letterbook, 1866–1883. Acc. No. 1056, Battye Library.
Governor's Despatches, 1850–1890, Western Australia. Acc. No. 390, Battye Library.
Lloyds Survey Register. National Maritime Museum, Greenwich.
MacIllroy, J. Dampier Archipeligo Historic Sites Survey, 1979. Report prepared for the Australian Heritage Commission.
New London Registry. New London, United States of America.
Police Department Records, Western Australia. Acc. No. 129, Battye Library.
Register of British Ships, Fremantle. Australia Shipping Registration Office, Department of Transport, Canberra.
Register of British Ships, Geelong. Australian Shipping Registration Office, Department of Transport, Canberra.
Register of British Ships, Hobart. Australian Shipping Registration Office, Department of Transport, Canberra.
Register of British Ships, Launceston. Australian Shipping Registration Office, Department of Transport, Canberra.
Register of British Ships, Melbourne. Australian Shipping Registration Office, Department of Transport, Canberra.
Register of British Ships, Port Adelaide. Australian Shipping Registration Office, Department of Transport, Canberra.
Register of British Ships, Sydney. Australian Archives, Sydney Regional Office.
Register of Shipping Arrivals and Departures at the Port of Fremantle, 1833–1881. Acc. No. 1076, Battye Library.

Shipping Register, Port Adelaide. Acc. No. SAA 436, Public Record Office of South Australia.

Sholl, R.J. Occurrence Book, Government Resident, Roebourne. Acc. 194, Battye Library.

Vickery, L. and Spooner, I. A Report on the 1980 Excavation of the *Hadda* site in Houtman Abrolhos. Western Australian Museum.

Western Australian Museum, Department of Maritime Archaeology Files. Western Australian Maritime Museum, Fremantle.

Private

Brockman, J. Diary, 1889. Acc. No. 1073A, Battye Library.

Cammilleri, C. Anthony Curtis: His Life in Western Australia. Typescript, 1963. Acc. No. B/Cur, Battye Library.

Clarke, H. Log Book of the Convict Ship *Stag* from Deptford to Western Australia. Acc. No. 276A, Battye Library.

Habgood Papers. Acc. No. 813A, Battye Library.

Henderson Papers. Norfolk Record Office, United Kingdom.

Heppingstone, I. Whaling at Port Gregory. Acc. No. PR7666, Battye Library.

Keesling Papers. Lt. Col. J. Keesling. In private hands.

Manning, L. Notes. Western Australian Museum.

Notes on the Dredger *Wallaroo* and the Hopper Barges *Kadina* and *Goolwa*. Acc. No. SAA RN88. Public Record Office of South Australia.

Samson Papers. Acc. Nos. 1120A and 2169A, Battye Library.

Scholl, T. Journal. Acc. No. QB/SHO, Battye Library.

Unpublished Theses

Henderson, G. From Sail to Steam: Shipping in Western Australia, 1870–1890. MA Thesis, University of Western Australia, 1977.

Hunt, S. The Gribble Affair. History Honours Degree Thesis, Murdoch University, 1978.

PUBLISHED SOURCES

Books

Bain, M. *Full Fathom Five*. Artlook Books, Perth, 1982.
Battye, J. *The History of the North West of Australia*. V. K. Jones, Perth, 1915.
Broeze, F. and Henderson, G. *Western Australians and the Sea: Our Maritime Heritage*. Western Australian Museum, Perth, 1986.
Burges, L. C. *The Pioneers of the Nor'-West, Australia*. Constantine and Gardner, Geraldton, 1913.
Carter, E. *Dictionary of Inventions and Discoveries*. Crane, Russak and Company, New York, 1976.
Crawford, I. *The Art of the Wandjina*. Oxford University Press, Melbourne, 1968.
Dakin, W. *Whalemen Adventurers*. Angus and Robertson, Sydney, 1938.
Fall, V. *The Sea and the Forest*. University of Western Australia Press, Nedlands, 1972.
Findlay, A. *A Directory for the Navigation of the Indian Ocean*. Laurie, London, 3rd edition 1876.
Fyfe, C. *The Bale Fillers: Western Australian Wool, 1826–1916*. University of Western Australia Press, Nedlands, 1983.
Goldsmith, F. *Treasure Lies Buried Here*. Pitman, Perth, 1946.
Grolier's *Australian Encyclopedia*, volume 10. Grolier Society, Sydney, 1965.
Henderson, G. *Unfinished Voyages: Western Australian Shipwrecks 1622–1850*. University of Western Australia Press, Nedlands, 1980.
Henderson, G. *Maritime Archaeology in Australia*. University of Western Australia Press, Nedlands, 1986.
Jarrett, F. *The Mercantile Navy List of Australia and New Zealand*. Loxton, Sydney, 1871.
Loney, J. *Australian Shipwrecks Volume 2, 1851–1871*. Reed, Terry Hills, New South Wales, 1979.
Loney, J. *Australian Shipwrecks Volume 3: 1871–1900*. List Publishing, Geelong, 1982.
Morley, F., and Hodgson, J. *Whaling North and South*. Methuen, London, 1927.
Reilly, J. *Reminiscences of Fifty Years of Residence in Western Australia*. Sands and McDougall, Perth, 1903.
Rhodes, F. *Pageant of the Pacific: Being the Maritime History of Australia*. Thwaites, Sydney, 1937.
Sanderson, I. *Follow the Whale*. Cassell, London, 1958.
Sledge, S. *Report of Wreck Inspection North Coast, 1978*. Western Australian Museum, Perth, 1980.

Somerville, W. *Rottnest Island, Its History and Legends*. Rottnest Island Board, Perth, 1966.

Starbuck, A. *History of the American Whale Fishery from its Earliest Inception to the Year 1876*. Argosy Antiquarian, New York, 1964.

Stow, J. P. *The Voyage of the Forlorn Hope, 1865*. Griffin Press for Sullivan's Cove, Adelaide, 1981.

Streeter, E. W. *Pearls and Pearling Life*. G. Bell and Sons, London, 1886.

Articles

Bray, F. 'George Clifton, 1823–1913'. *The Royal Western Australian Historical Society Journal and Proceedings*, 2: 20, 1936, pp. 1–25.

Broeze, F. 'Western Australia until 1869: The Maritime Perspective.' *Early Days: Journal of the Royal Western Australian Historical Society*, 8(5 & 6), 1981 & 1982

Drake-Brockman, H. 'Charles Edward Broadhurst'. *Australian Dictionary of Biography*, Volume 3, 1969, pp. 293–304.

Gara, T. 'The Flying Foam Massacre: An Incident on the North West Frontier, Western Australia', Smith, M. (ed.), *Archaeology at ANZAAS, 1983*. Western Australian Museum, Perth, 1983, pp. 86–94.

Hicks, A. 'First into the Kimberleys'. *Early Days: Journal of the Royal Western Australian Historical Society*, 3, October 1938, pp. 11–19.

McCarthy, M. 'SS *Xantho*: A 19th Century Lemon Turned Sweet'. *Historical Archaeology Special Publication No. 4*, 1985, pp. 54–6.

Parsons, R. 'The Pearling Fleet in Western Australia', Kerr, G. (ed.), *Australian and New Zealand Sail Traders*. Lynton Publications, Blackwood, South Australia, 1974, pp. 11–36.

Shorten, Ann, P. 'To Take a Star . . . A Review of Australian Statutory Qualifications for Merchant Seafarers. 1852–1869'. *The Great Circle*, 1: 1, 1979, pp. 15–32.

Smoje, N. 'Shipwrecked on the North West Coast: The Ordeal of the Survivors of the *Stefano*'. *Early Days: Journal of the Royal Western Australian Historical Society*, 8: 2, 1978, pp. 35–50.

Stevens, G. 'The East-West Telegraph, 1875–77'. *The Royal Western Australian Historical Society Journal and Proceedings*, 2: 13, 1933, p. 16–35.

Summerville, H. 'Port Gregory'. *Early Days: Journal of the Royal Western Australian Historical Society*, 6: 8, 1969, pp. 74–88.

Newspapers (showing dates consulted)

Albany Advertiser, 1897. Albany.
Commercial News and Shipping List, 1855. Perth.
Countryman, 1955. Perth.
Handelsblad, 1880. Batavia.
Herald, 1867–1886. Perth.
Inquirer, 1840–1855. Perth.
Inquirer and Commercial News, 1855–1890. Perth.
Melbourne Argus, 1853. Melbourne.
Perth Gazette, 1833–1864. Perth.
Perth Gazette and W.A. Times, 1864–1874. Perth.
South Australian Register, 1836–1931. Adelaide.
Whaleman's Shipping List and Merchants Transcript, 1856. Boston.
Western Australian Times, 1863–1864. Perth.
Western Australian Times, 1874–1879. Perth.
West Australian, 1879– . Perth.

Periodicals (showing dates consulted)

Australian Shipping Register, 1968. Adelaide.
Blue Books, 1851–1880. Perth.
Government Gazette, 1851–1880. Perth.
Lloyds Shipping Register, 1850–1890. London.
Mercantile Navy List, 1882–1890. London.
Muresk College Magazine, September 1928. Northam.
Registre Veritas, 1829–1890. Clearwater Publishing Co, New York, 1978.
South Australian Register, 1856. Adelaide.
Underwater Explorers Club News, 1963– . Perth.
Western Australian Almanac, 1849–1874. Perth.
Western Australian Calendar and Directory, 1853–1880.

Indexes

Index of Vessels

NAMED VESSELS

UNNAMED VESSELS

Index of Persons and Organizations

Index of Places